Natural Computing Series

Series Editors: G. Rozenberg
Th. Bäck A.E. Eiben J.N. Kok H.P. Spaink

Leiden Center for Natural Computing

Springer
Berlin
Heidelberg
New York
Hong Kong
London
Milan
Paris
Tokyo

Mika Hirvensalo

Quantum Computing

Second Edition

Springer

Mika Hirvensalo

University of Turku
Department of Mathematics
20014 Turku
Finland

mikhirve@utu.fi

Series Editors

G. Rozenberg (Managing Editor)
rozenber@liacs.nl
Th. Bäck, J. N. Kok, H. P. Spaink

Leiden Center for Natural Computing, Leiden University
Niels Bohrweg 1, 2333 CA Leiden, The Netherlands

A. E. Eiben
Vrije Universiteit Amsterdam

Library of Congress Cataloging-in-Publication Data
Hirvensalo, Mika, 1972–
Quantum computing / M. Hirvensalo.
p. cm. – (Natural computing series)
Includes bibliographical references and index.

1. Quantum computers. I. Title. II. Series.
QA76.889.H57 2003 004.1–dc22 2003066405

ACM Computing Classification (1998): F.1-2, G.1.2, G3, H.1.1, I.1.2, J.2

ISBN 978-3-642-07383-0

Springer-Verlag is a part of Springer Science+Business Media
springeronline.com

Printed in Germany

Cover design: KünkelLopka, Heidelberg

Printed on acid-free paper 45/3142PS – 5 4 3 2 1 0

Preface to the Second Edition

After the first edition of this book was published, I received much positive feedback from the readers. It was very helpful to have all those comments suggesting improvements and corrections. In many cases, it was suggested that more aspects on quantum information would be welcome. Unfortunately, I am afraid that an attempt to cover such a broad area as quantum information theory would make this book too scattered to be helpful for educational purposes.

On the other hand, I admit that some aspects of quantum information should be discussed. The first edition already contained the so-called No-Cloning Theorem. In this edition, I have added a stronger version of the aforementioned theorem due to R. Jozsa, a variant which also covers the no-deleting principle. Moreover, in this edition, I have added some famous protocols, such as *quantum teleportation.*

The response to the first edition strongly supports the idea that the main function of this book should be educational, and I have not included further aspects of quantum information theory here. For further reading, I suggest [43] by Josef Gruska and [62] by Michael A. Nielsen and Isaac L. Chuang.

Chapter 1, especially Section 1.4, includes the most basic knowledge for the presentation of quantum systems relevant to quantum computation.

The basic properties of quantum information are introduced in Chapter 2. This chapter also includes interesting protocols: quantum teleportation and superdense coding.

Chapter 3 is divided as follows: Turing machines, as well as their probabilistic counterparts, are introduced in Section 3.1 as traditional, *uniform* models of computation. For the reader interested in quantum computing but having little knowledge of the theory of computation, this section was designed to also include the basic definitions of complexity theory. Section 3.1 is intended for the reader who has a solid background in quantum mechanics, but little previous knowledge on classical computation theory. The reader who is well aware of the theory of computation may skip Section 3.1: for such a reader the knowledge in Chapter 2 and Section 3.2 (excluding the first subsection) is sufficient to follow this book. In Section 3.2, we represent Boolean and quantum circuits (as an extension of the concept of reversible circuits) as models of computation. Because of their descriptional simplicity, circuits are used throughout this book to present quantum algorithms.

Chapter 4 is devoted to a complete representation of Shor's famous factorization algorithm. The instructions in Chapter 4 will help the reader to choose which sections may, according to the reader's background, be skipped. Chapter 5 is closely connected to Chapter 4, and can be seen as a representation of a structural but not straightforward extension of Shor's algorithm.

Chapter 6 was written to introduce Grover's method for obtaining a quadratic speedup in quantum computation (with respect to classical computation) in its basic form, whereas in Chapter 7 we represent a method for obtaining lower bounds for quantum computation in a restricted quantum circuit model.

Chapters 8 and 9 are appendices intended for the beginner, but Chapter 8 is also suitable for the reader who has a strong background in computer science and is interested in quantum computation. Chapter 9 is composed of various different topics in mathematics, since it has already turned out that, in the area of quantum computation, many mathematical disciplines, seemingly separate from each other, are useful. Moreover, my personal experience is that a basic education in computer science and physics very seldom covers all the areas in Chapt. 9.

Acknowledgements. That this second edition ever emerged is due to Ingeborg Mayer. I wish to thank her very much for her complaisant way of cooperating. Thanks also go to all the readers of the first edition who sent me suggestions about corrections and improvements.

Turku, Finland
October 2003 Mika Hirvensalo

From the Preface to the First Edition

The twentieth century witnessed the birth of revolutionary ideas in the physical sciences. These ideas began to shake the traditional view of the universe dating back to the days of Newton, even to the days of Galileo. Albert Einstein is usually identified as the creator of the *relativity theory*, a theory that is used to model the behavior of the huge macrosystems of astronomy. Another new view of the physical world was supplied by *quantum physics*, which turned out to be successful in describing phenomena in the microworld, the behavior of particles of atomic size.

Even though the first ideas of automatic information processing are quite old, I feel justified in saying that the twentieth century also witnessed the birth of computer science. As a mathematician, by the term "computer science", I mean the more theoretical parts of this vast research area, such as the theory of formal languages, automata theory, complexity theory, and algorithm design. I hope that readers who are used to a more flexible concept of "computer science" will forgive me. The idea of a computational device was crystallized into a mathematical form as a *Turing machine* by Alan Turing in the 1930s. Since then, the growth of computer science has been immense, but many problems in newer areas such as complexity theory are still waiting for a solution.

Since the very first electronic computers were built, computer technology has grown rapidly. An observation by Gordon Moore in 1965 laid the foundations for what became known as "Moore's Law" – that computer processing power doubles every eighteen months. How far can this technical process go? How efficient can we make computers? In light of the present knowledge, it seems unfair even to attempt to give an answer to these questions, but some estimate can be given. By naively extrapolating Moore's law to the future, we learn that sooner or later, each bit of information should be encoded by a physical system of subatomic size! Several decades ago such an idea would have seemed somewhat absurd, but it does not seem so anymore. In fact, a system of seven bits encoded subatomically has been already implemented [51]. These small systems can no longer be described by classical physics, but rather quantum physical effects must be taken into consideration.

When thinking again about the formalization of a computer as a Turing machine, rewriting system, or some other classical model of computation, one realizes that the concept of *information* is usually based on strings over a finite alphabet. This strongly reflects the idea of classical physics in the following sense: each member of a string can be represented by a physical system (storing the members in the memory of an electronic computer, writing them on sand, etc.) that can be in a certain *state*; i.e., contain a character of the alphabet. Moreover, we should be able to identify different states reliably. That is, we should be able to make an *observation* in such a way that we become convinced that the system under observation represents a certain character.

In this book, we typically identify the alphabet and the distinguishable states of a physical system that represent the information. These identifiable states are called *basis states*. In quantum physical microsystems, there are also basis states that can be identified and, therefore, we could use such microsystems to represent information. But, unlike the systems of classical physics, these microsystems are also able to exist in a *superposition* of basis states, which, informally speaking, means that the state of such a system can also be a combination of basis states. We will call the information represented by such microsystems *quantum information*. One may argue that in classical physics it is also possible to speak about combinations of basis states: we can prepare a *mixed* state which is essentially a probability distribution of the basis states. But there is a difference between the superpositions of quantum physics and the probability distributions of classical physics: due to the *interference effects*, the superpositions cannot be interpreted as mixtures (probability distributions) of the basis states.

Richard Feynman [38] pointed out in 1982 that it appears to be extremely difficult by using an ordinary computer to simulate efficiently how a quantum physical system evolves in time. He also demonstrated that, if we had a computer that runs according to the laws of quantum physics, then this simulation could be made efficiently. Thus, he actually suggested that a *quantum computer* could be essentially more efficient than any traditional one.

Therefore, it is an interesting challenge to study *quantum computation*, the theory of computation in which traditional information is replaced by its quantum physical counterpart. Are quantum computers more powerful than traditional ones? If so, what are the problems that can be solved more efficiently by using a quantum computer? These questions are still waiting for answers.

The purpose of this book is to provide a good introduction to quantum computation for beginners, as well as a clear presentation of the most important presently known results for more advanced readers. The latter purpose also includes providing a bridge (from a mathematician's point of view) between quantum mechanics and the theory of computation: it is not only my personal experience that the language used in research articles on these topics is completely different.

This book is concentrated mainly on quantum algorithms but other interesting topics, such as quantum information theory, quantum communication, quantum error-correcting, and quantum cryptography, are not covered. Therefore, for additional reading, we can warmly recommend [43] by Josef Gruska. Book [43] also contains a large number of references to works on quantum computing. A reader who is more oriented to physics may also see [89] and [90] by C. P. Williams and S. H. Clearwater. It may also be useful to follow the Los Alamos preprint archive at http://xxx.lanl.gov/archive/quant-ph to learn about the new developments in quantum computing. The Los Alamos preprint archive contains a large number of articles on quantum computing since 1994, and includes many articles referred to in this book.

Acknowledgements. My warmest thanks go to Professors Grzegorz Rozenberg and Arto Salomaa for encouraging me to write this book, and to Professor Juhani Karhumäki and Turku Centre for Computer Science for providing excellent working conditions during the writing period. This work has been supported by the Academy of Finland under grants 14047 and 44087. I am also indebted to Docent P. J. Lahti, V. Halava, and M. Rönkä for their careful revision work on parts of this book. Thanks also go to Dr. Hans Wössner for a fruitful cooperation.

Turku, Finland, February 2001 *Mika Hirvensalo*

Contents

1. Introduction

1.1 A Brief History of Quantum Computation

In connection to computational complexity, it could be stated that the theory of quantum computation was launched in the beginning of the 1980s. A most famous physicist, Nobel Prize winner Richard P. Feynman, proposed in his article [38] which appeared in 1982 that a quantum physical system of R particles cannot be simulated by an ordinary computer without an *exponential* slowdown in the efficiency of the simulation. However, a system of R particles in classical physics can be simulated well with only a polynomial slowdown. The main reason for this is that the *description size* of a particle system is linear in R in classical physics,[1] but exponential in R according to quantum physics (In Section 1.4 we will learn about the quantum physics description). Feynman himself expressed:

> But the full description of quantum mechanics for a large system with R particles is given by a function $\psi(x_1, x_2, \ldots, x_R, t)$ which we call the amplitude to find the particles $x_1, \ldots, x_R$, and therefore, because it has too many variables, it *cannot be simulated* with a normal computer with a number of elements proportional to R or proportional to N. [38]

Feynman also suggested that this slowdown could be avoided by using a computer running according to the laws of quantum physics. This idea suggests, at least implicitly, that a *quantum computer* could operate exponentially faster than a deterministic classical one. In [38], Feynman also addressed the problem of simulating a quantum physical system with a *probabilistic computer* but due to interference phenomena, it appears to be a difficult problem.

Quantum mechanical computation models were also constructed by Benioff [7] in 1982, but Deutsch argued in [31] that Benioff's model can be perfectly simulated by an ordinary computer. In 1985 in his notable paper [31], Deutsch was the first to establish a solid ground for the theory of quantum computation by introducing a fully quantum model for computation and giving the description of a *universal quantum computer*. Later, Deutsch also defined quantum networks in [32]. The construction of a *universal quantum Turing machine* was improved by Bernstein and Vazirani in [13], where the

[1] One should give the coordinates and the momentum of each particle with required precision.

authors show how to construct a universal quantum Turing machine capable of simulating any other quantum Turing machine with *polynomial efficiency.*

After the pioneering work of D. Deutsch, quantum computation still remained a marginal curiosity in the theory of computation until 1994, when Peter W. Shor introduced his celebrated quantum algorithms for factoring integers and extracting discrete logarithms in polynomial time [81]. The importance of Shor's algorithm for finding factors is well-known: reliability of the famous RSA cryptosystem designed for secret communications is based on the assumption that the factoring of large integers will remain an intractable problem, but Shor demonstrated that this is not true if one could build a quantum computer.

However, the theory is far more developed than the practice: a large-scale quantum computer has not been built yet. A major difficulty arises from two somehow contradictory requirements. On the one hand the computer memory consisting of a microscopically small quantum system must be isolated as perfectly as possible to prevent destructive interaction with the environment. On the other hand, the "quantum processing unit" cannot be totally isolated, since the computation must carry on, and a "supervisor" should ensure that the quantum system evolves in the desired way. In principle, the problem that non-controllable errors occur is not a new one: in classical information theory we consider a noisy channel that may corrupt the messages. The task of the receiver is to extract the correct information from the distorted message *without any additional information transmission.* The classical theory of error-correcting codes [57] addresses this problem, and the elementary result of this theory originally, due to Claude Shannon, could be expressed as follows: for a reasonably erratic channel, there is a coding system of messages which allows us to decrease the error probability in the transmission as much as we want.

It was first believed that a corresponding scheme for quantum computations was impossible even in theory, mainly because of the No-cloning Theorem [91], which states that quantum information cannot be duplicated exactly. However, Shor demonstrated in his article [82] how we can construct error-correcting schemes for quantum computers, thus establishing the theory of *quantum error-correcting codes.* To learn more about this theory, the reader may consult [24], for example. In article [72], J. Preskill emphasizes that quantum error-correcting codes may someday lead to the construction of a large-scale quantum computer, but this remains to be seen.

1.2 Classical Physics

Physics, as we can understand it today, is the theory of overall nature. This theory is naturally too broad to be profoundly accessed in a brief moment, so we are satisfied just to point out some essential features that will be important when learning about the differences between quantum and classical computation.

As its very core, physics is ultimately an empirical science in the sense that a physical theory can be regarded valid only if the theory agrees with empirical observations. Therefore, it is not surprising that the concept of *observables*[2] has great importance in the physical sciences. There are observables associated with a physical system, like *position* and *momentum*, to mention a few. The description of a physical system is called the *state* of the system.

Example 1.2.1. Assume that we would like to describe the mechanics of a single particle X in a closed region of space. The observables used for the system description are the position and the momentum. Thus, we can, under a fixed coordinate system, express the state of the system as a vector $\boldsymbol{x} = (x_1, x_2, x_3, p_1, p_2, p_3) \in \mathbb{R}^6$, where (x_1, x_2, x_3) describes the position and (p_1, p_2, p_3) the momentum.

As the particle moves, the state of the system changes in time. The way in which classical mechanics describes the *time evolution* of the state is given by the *Hamiltonian equations of motion*:

$$\frac{\mathrm{d}}{\mathrm{d}t}x_i = \frac{\partial}{\partial p_i}H, \qquad \frac{\mathrm{d}}{\mathrm{d}t}p_i = -\frac{\partial}{\partial x_i}H,$$

where $H = H(x_1, x_2, x_3, p_1, p_2, p_3)$ is an observable called the *Hamiltonian function* of the system.

Suppose now that another particle Y is inserted into our system. Then the full description of the system is given as a vector $(\boldsymbol{x}, \boldsymbol{y}) \in \mathbb{R}^6 \times \mathbb{R}^6$, where $\boldsymbol{x}$ is as before, and $\boldsymbol{y}$ describes the position and the momentum of the second particle.

To build the mathematical description of a physical system, it is sufficient to assume that all the observables take their values in a set whose cardinality does not exceed the cardinality of the real number system. Therefore, we can assume that the observables take real number values; if it is natural to think that a certain observable takes values in some other set A, we can always replace the elements of A with suitable real numbers. Keeping this agreement in mind, we can list the properties of a description of a physical system.

- A physical system is described by a *state vector* (also called *state*) $\boldsymbol{x} \in \mathbb{R}^k$ for some k. The set of states is called the *phase space* of the system. The state is the full description of the dynamic properties of interest. This means that the state completely describes the properties whose development we actually want to describe. The other physical properties, such as electrical charge, temperature, etc., are not included in the state if they do

[2] In classical mechanics, observables are usually called *dynamic variables*.

not alter or do not affect the properties of interest. It should be emphasized here that, in the previous example, we were interested in describing the mechanics of a particle, so the properties of interest are its position and momentum. However, it may be argued that if we investigate a particle in an electrical field, the electrical charge of the particle definitely affects its behaviour: the greater the charge, the greater the acceleration. But in this case, the charge is seen as a constant property of the system and is not encoded in the state.

- The state depends on time, so instead of $\boldsymbol{x}$ we should actually write $\boldsymbol{x}(t)$ or $\boldsymbol{x}_t$. If the system is regular enough, as classical mechanics is, a state also determines the future states (and also the past states). That is, we can find a time-dependency which is a function U_t such that $\boldsymbol{x}(t) = U_t(\boldsymbol{x}(0))$. This U_t, of course, depends on the system itself, as well as on the constant properties of the system. In our first example, U_t is determined by the Hamiltonian via the Newtonian equations of motion.
- If two systems are described by states $\boldsymbol{x}$ and $\boldsymbol{y}$, the state of the compound system consisting of both systems can be written as $(\boldsymbol{x}, \boldsymbol{y})$. That is, the description of the compound system is a Cartesian product of the subsystems.

1.3 Probabilistic Systems

In order to be able to comment on some characteristic features of the representation of quantum systems, we first study the representation of a *probabilistic system.* A system admitting probabilistic nature means that we do not know *for certain* the state of the system, but we do know the *probability distribution* of the states. In other words, we know that system is in states $\boldsymbol{x}_1, \ldots, \boldsymbol{x}_n$ with probabilities $p_1, \ldots, p_n$ that sum up to 1 (if there is a continuum of the possible states as in Example 1.2.1, we should have a probability distribution that integrates up to 1, but for the sake of simplicity, we will study here only systems having finitely many states). Notation

$$p_1[\boldsymbol{x}_1] + p_2[\boldsymbol{x}_2] + \ldots + p_n[\boldsymbol{x}_n], \tag{1.1}$$

where $p_i \geq 0$ and $p_1 + \ldots + p_n = 1$ stands for a probability distribution, meaning that the system is in state $\boldsymbol{x}_i$ with probability p_i. We also call distribution (1.1) a *mixed state.* Hereafter, states $\boldsymbol{x}_i$ are called *pure states.*

Now make a careful distinction between the notations: Expression (1.1) *does not* mean the *expectation value* (also called the *average*) of the state

$$p_1\boldsymbol{x}_1 + p_2\boldsymbol{x}_2 + \ldots + p_n\boldsymbol{x}_n \in \mathbb{R}^k, \tag{1.2}$$

but (1.1) is only the probability distribution over states $\boldsymbol{x}_i$.

Example 1.3.1. Tossing a fair coin will give head $\boldsymbol{h}$ or tail $\boldsymbol{t}$ with a probability of $\frac{1}{2}$. According to classical mechanics, we may think that, in principle, perfect knowledge about the coin and all circumstances connected to the tossing will allow us to determine the outcome with certainty. However, in practice it is impossible to handle all these circumstances, and the notation $\frac{1}{2}[\boldsymbol{h}] + \frac{1}{2}[\boldsymbol{t}]$ for the mixed state of a fair coin reflects our lack of information about the system.

Using an auxiliary system, we can easily introduce a probabilistic nature for any system. In fact, this is what we usually do in connection with probabilistic algorithms: the algorithm itself is typically thought to be strictly deterministic, but it utilizes *random bits*, which are supposed to be freely available.[3]

Example 1.3.2. Let us assume that the time evolution of a system with pure states $\boldsymbol{x}_1, \ldots, \boldsymbol{x}_n$ also depends on an auxiliary system with pure states $\boldsymbol{h}$ and $\boldsymbol{t}$, such that the compound system state $(\boldsymbol{x}_i, \boldsymbol{h})$ evolves during a fixed time interval into $(\boldsymbol{x}_{h(i)}, \boldsymbol{h})$ and $(\boldsymbol{x}_i, \boldsymbol{t})$ into $(\boldsymbol{x}_{t(i)}, \boldsymbol{t})$, where $h, t : \{1, \ldots, n\} \mapsto \{1, \ldots, n\}$ are some functions. The auxiliary system with states $\boldsymbol{h}$ and $\boldsymbol{t}$ can thus be interpreted as a *control system* which indicates how the original one should behave.

Let us then consider the control system in a mixed state $p_1[\boldsymbol{h}] + p_2[\boldsymbol{t}]$ (if $p_1, p_2 \neq \frac{1}{2}$, we may call the control system *a biased coin*). The compound state can then be written as a mixture

$$p_1[(\boldsymbol{x}_i, \boldsymbol{h})] + p_2[(\boldsymbol{x}_i, \boldsymbol{t})],$$

which evolves into a mixed state

$$p_1[(\boldsymbol{x}_{h(i)}, \boldsymbol{h})] + p_2[(\boldsymbol{x}_{t(i)}, \boldsymbol{t})].$$

If the auxiliary system no longer interferes with the first one, we can ignore it and write the state of the first system as

$$p_1[\boldsymbol{x}_{h(i)}] + p_2[\boldsymbol{x}_{t(i)}].$$

A control system in a mixed state is called a *randomizer*. Supposing that the randomizers are always available, we may even assume that the system under consideration evolves probabilistically and ignore the randomizer.

Using a more general concept of randomizer, we can achieve notational and conceptual simplification by assuming that the time evolution of a system is not a deterministic procedure, but develops each state $\boldsymbol{x}_i$ into a distribution

$$\boldsymbol{x}_i \mapsto p_{1i}[\boldsymbol{x}_1] + p_{2i}[\boldsymbol{x}_2] + \ldots + p_{ni}[\boldsymbol{x}_n], \tag{1.3}$$

[3] In classical computation, generating random bits is a very complicated issue. For further discussion on this topic, consult section 11.3 of [64].

such that $p_{1i}+p_{2i}+\ldots+p_{ni}=1$ for each i. In (1.3), p_{ji} is the probability that the system state $\boldsymbol{x}_i$ evolves into $\boldsymbol{x}_j$. Notice that we have now also made the time discrete in order to simplify the mathematical framework, and that this actually is well-suited to the computational aspects: we can assume that we have instantaneous descriptions of the system between short time intervals, and that during each interval, the system undergoes the time evolution (1.3). Of course, the time evolution does not always need to be the same, but rather may depend on the particular interval. Considering this probabilistic time evolution, the notation (1.1) appears to be very handy: during a time interval, a distribution

$$p_1[\boldsymbol{x}_1]+p_2[\boldsymbol{x}_2]+\ldots+p_n[\boldsymbol{x}_n] \tag{1.4}$$

evolves into

$$\begin{aligned} & p_1(p_{11}[\boldsymbol{x}_1]+\ldots+p_{n1}[\boldsymbol{x}_n])+\ldots+p_n(p_{1n}[\boldsymbol{x}_1]+\ldots+p_{nn}[\boldsymbol{x}_n]) \\ = & (p_{11}p_1+\ldots+p_{1n}p_n)[\boldsymbol{x}_1]+\ldots+(p_{n1}p_1+\ldots+p_{nn}p_n)[\boldsymbol{x}_n] \\ = & p_1'[\boldsymbol{x}_1]+p_2'[\boldsymbol{x}_2]+\ldots+p_n'[\boldsymbol{x}_n], \end{aligned}$$

where we have denoted $p_i'=p_{i1}p_1+\ldots+p_{in}p_n$. The probabilities p_i and p_i' are thus related by

$$\begin{pmatrix} p_1' \\ p_2' \\ \vdots \\ p_n' \end{pmatrix} = \begin{pmatrix} p_{11} & p_{12} & \cdots & p_{1n} \\ p_{21} & p_{22} & \cdots & p_{2n} \\ \vdots & \vdots & \ddots & \vdots \\ p_{n1} & p_{n2} & \cdots & p_{nn} \end{pmatrix} \begin{pmatrix} p_1 \\ p_2 \\ \vdots \\ p_n \end{pmatrix}. \tag{1.5}$$

Notice that the matrix in (1.5) has non-negative entries and $p_{1i}+p_{2i}+\ldots+p_{ni}=1$ for each i, which guarantees that $p_1'+p_2'+\ldots+p_n'=p_1+p_2+\ldots+p_n$. A matrix with this property is called a *Markov matrix*. A probabilistic system with a time evolution described above is called a *Markov chain*.

Notice that a distribution (1.4) can be considered as a vector with non-negative coordinates that sum up to 1 in an n-dimensional real vector space having $[\boldsymbol{x}_1]$, ..., $[\boldsymbol{x}_n]$ as basis vectors. The set of all mixed states (distributions) is a convex set,[4] having the pure states as extremals. Unlike in the representation of quantum systems, the fact that the mixed states are elements of a vector space is not very important. However, it may be convenient to describe the probabilistic time evolution as a Markov mapping, i.e., a linear mapping which preserves the property that the coordinates are non-negative and sum up to 1. For an introduction to Markov chains, see [75], for instance.

[4] A set S of vectors is *convex*, if for all $\boldsymbol{x}_1$, $\boldsymbol{x}_2 \in S$ and all $p_1, p_2 \geq 0$ such that $p_1+p_2=1$ also $p_1\boldsymbol{x}_1+p_2\boldsymbol{x}_2 \in S$. An element $\boldsymbol{x} \in S$ of a convex set is an *extremal*, if $\boldsymbol{x}=p_1\boldsymbol{x}_1+p_2\boldsymbol{x}_2$ implies that either $p_1=0$ or $p_2=0$.

1.4 Quantum Mechanics

In the beginning of the twentieth century, experiments on atoms and radiation physics gave strong support to the idea that there are physical systems that cannot be satisfactorily described even by using the Markov chain representation of the previous section. In Section 8.1 we will discuss these experiments in more detail, but here in the introductory chapter we are satisfied with only presenting the mathematical description of a quantum mechanical system. We would like to emphasize that this representation based on so-called *state vectors* which is studied here is not the most general representation of quantum systems. A general *Hilbert space formalism* of quantum mechanics is represented in Section 8.4. The advantage of a representation using only state vectors is that it is mathematically simpler than the more general one. To explain the terminology, we usually use "quantum mechanics" to mean the mathematical structure that describes "quantum physics", which can be understood more generally.

The quantum mechanical description of a physical system looks very much like the probabilistic representation

$$p_1[\boldsymbol{x}_1] + p_2[\boldsymbol{x}_2] + \ldots + p_n[\boldsymbol{x}_n], \tag{1.6}$$

but still differs essentially from (1.6).

In fact, in quantum mechanics a state of an *n-level system* is depicted as a unit-length vector in an n-dimensional complex vector space H_n (see Section 9.3).[5] We call H_n the *state space* of the system. To illustrate, let us choose an orthonormal basis $\{|\boldsymbol{x}_1\rangle, \ldots, |\boldsymbol{x}_n\rangle\}$ for the state space H_n (we assume here that n is a finite number). This strange "ket"-notation $|\boldsymbol{x}\rangle$ is originally due to P. Dirac, and its usefulness will become apparent later. Now, any state of our quantum system can then be written as

$$\alpha_1 |\boldsymbol{x}_1\rangle + \alpha_1 |\boldsymbol{x}_2\rangle + \ldots + \alpha_n |\boldsymbol{x}_n\rangle, \tag{1.7}$$

where α_i are complex numbers called the *amplitudes* (with respect to the chosen basis) and the requirement of unit length means that $|\alpha_1|^2 + |\alpha_2|^2 + \ldots + |\alpha_n|^2 = 1$.

It should be immediately emphasized that the choice of the orthonormal basis $\{|\boldsymbol{x}_1\rangle, \ldots, |\boldsymbol{x}_n\rangle\}$ is arbitrary but, for any such fixed basis, refers to a physical observable which can take n values. To simplify the framework, we do not associate any numerical values with the observables in this section, but we merely say that the system can have properties $\boldsymbol{x}_1, \ldots, \boldsymbol{x}_n$. Numerical values associated to the observables are handled in Section 8.3.2. The amplitudes α_i induce a probability distribution in the following way: the probability that a system in a state (1.7) is seen to have property $\boldsymbol{x}_i$ is $|\alpha_i|^2$ (we also say that the

[5] A finite-dimensional vector space over complex numbers is an example of a *Hilbert space*.

probability that $\boldsymbol{x}_i$ is *observed* is $|\alpha_i|^2$). The basis vectors $|\boldsymbol{x}_i\rangle$ are called the *basis states* and (1.7) is referred to as a *superposition of basis states.* Mapping $\psi(\boldsymbol{x}_i) = \alpha_i$ is called the *wave function* with respect to basis $|\boldsymbol{x}_1\rangle, \ldots, |\boldsymbol{x}_n\rangle$.

Even though this state vector formalism is simpler than the general one, it has one inconvenient feature: if $|\boldsymbol{x}\rangle, |\boldsymbol{y}\rangle \in H_n$ are any states that satisfy $|\boldsymbol{x}\rangle = e^{i\theta}|\boldsymbol{y}\rangle$, we say that states $|\boldsymbol{x}\rangle$ and $|\boldsymbol{y}\rangle$ are *equivalent.* Clearly equivalent states induce the same probability distribution over basis states (with respect to any chosen basis), and we usually identify equivalent states.

Example 1.4.1. A two-level quantum mechanical system can be used to represent a bit, and such a system is called a *quantum bit.* For such a system we choose an orthonormal basis $\{|0\rangle, |1\rangle\}$, and a general state of the system is

$$\alpha_0 |0\rangle + \alpha_1 |1\rangle,$$

where $|\alpha_0|^2 + |\alpha_1|^2 = 1$. The above superposition induces a probability distribution such that the probabilities that the system is seen to have properties 0 and 1 are $|\alpha_0|^2$ and $|\alpha_1|^2$ respectively.

What about the time evolution of quantum systems? The time evolution of a probabilistic system via Markov matrices is replaced by matrices with complex number entries that preserve quantity $|\alpha_1|^2 + \ldots + |\alpha_n|^2$. Thus, the quantum system in a state

$$\alpha_1 |\boldsymbol{x}_1\rangle + \ldots + \alpha_n |\boldsymbol{x}_n\rangle$$

evolves during a time interval into the state

$$\alpha_1' |\boldsymbol{x}_1\rangle + \ldots + \alpha_n' |\boldsymbol{x}_n\rangle,$$

where amplitudes $\alpha_1, \ldots, \alpha_n$ and $\alpha_1', \ldots, \alpha_n'$ are related by

$$\begin{pmatrix} \alpha_1' \\ \alpha_2' \\ \vdots \\ \alpha_n' \end{pmatrix} = \begin{pmatrix} a_{11} & a_{12} & \cdots & a_{1n} \\ a_{21} & a_{22} & \cdots & a_{2n} \\ \vdots & \vdots & \ddots & \vdots \\ a_{n1} & a_{n2} & \cdots & a_{nn} \end{pmatrix} \begin{pmatrix} \alpha_1 \\ \alpha_2 \\ \vdots \\ \alpha_n \end{pmatrix}, \tag{1.8}$$

and $|\alpha_1'|^2 + \ldots + |\alpha_n'|^2 = |\alpha_1|^2 + \ldots + |\alpha_n|^2$. It turns out that the matrices satisfying this requirement are *unitary matrices* (see Section 8.3.1 for details). Unitarity of a matrix A means that the transpose complex conjugate of A, denoted by A^*, is the inverse matrix to A. Matrix A^* is also called the *adjoint matrix*[6] of A. By saying that the time evolution of quantum systems is unitary, several authors mean that the evolution of quantum systems are determined by unitary matrices. Notice that unitary time evolution has very

[6] Among physics-oriented authors, there is also a widespread tradition of denoting the adjoint matrix by $A^\dagger$.

interesting consequences: unitarity especially means that the time evolution of a quantum system is *invertible*. If fact, $(\alpha_1, \ldots, \alpha_n)$ can be perfectly recovered from $(\alpha'_1, \ldots, \alpha'_n)$, since the matrix in (1.8) has an inverse.

We interrupt the description of quantum systems for a moment to discuss the differences between probabilistic and quantum systems. As we have said before, a mixed state (probability distribution)

$$p_1[\boldsymbol{x}_1] + \ldots + p_n[\boldsymbol{x}_n] \tag{1.9}$$

of a probabilistic system and a superposition

$$\alpha_1 |\boldsymbol{x}_1\rangle + \ldots + \alpha_n |\boldsymbol{x}_n\rangle \tag{1.10}$$

of a quantum system formally resemble each other very closely, and therefore it is quite natural if (1.10) can be seen as a generalization of (1.9). At first glance, the interpretation that (1.10) induces the probability distribution such that $|\alpha_i|^2$ is the probability of observing $\boldsymbol{x}_i$ may only seem like a technical difference. Can (1.10) also be interpreted to represent our ignorance such that the system in state (1.10) is actually in some state $|\boldsymbol{x}_i\rangle$ with a probability of $|\alpha_i|^2$? The answer is absolutely **no**. A fundamental difference can be found even by recalling that the orthonormal basis of H_n can be chosen freely. We can, in fact, choose an orthonormal basis $|\boldsymbol{x}'_1\rangle$, ..., $|\boldsymbol{x}'_n\rangle$ such that

$$|\boldsymbol{x}'_1\rangle = \alpha_1 |\boldsymbol{x}_1\rangle + \ldots + \alpha_n |\boldsymbol{x}_n\rangle ,$$

so, with respect to the new basis, the state of the system is simply $|\boldsymbol{x}'_1\rangle$. The new basis refers to another physical observable, and with respect to this observable, the system may have some of the properties $\boldsymbol{x}'_1$, ..., $\boldsymbol{x}'_n$. But in the state $|\boldsymbol{x}'_1\rangle$, the system is seen to have property $\boldsymbol{x}'_1$ with a probability of 1.

Example 1.4.2. Consider a quantum system with two basis states $\boldsymbol{t}$ and $\boldsymbol{h}$ (of course this system is the same as a quantum bit, only the terminology is different for illustration). Here we call this system a *quantum coin*. We consider a time evolution

$$|\boldsymbol{h}\rangle \mapsto \frac{1}{\sqrt{2}} |\boldsymbol{h}\rangle + \frac{1}{\sqrt{2}} |\boldsymbol{t}\rangle$$
$$|\boldsymbol{t}\rangle \mapsto \frac{1}{\sqrt{2}} |\boldsymbol{h}\rangle - \frac{1}{\sqrt{2}} |\boldsymbol{t}\rangle ,$$

which we here call a *fair coin toss* (verify that the time evolution is unitary). Notice that, beginning with either state $|\boldsymbol{h}\rangle$ or $|\boldsymbol{t}\rangle$, the state after the toss is one of the above states on the right-hand side, and that both of them have the property that $\boldsymbol{h}$ and $\boldsymbol{t}$ will both be observed with a probability of $\frac{1}{2}$. Imagine then that we begin with state $|\boldsymbol{h}\rangle$ and perform the fair coin toss *twice*. After the first toss, the state is as above, but if we do not observe the system, the state will be $|\boldsymbol{h}\rangle$ again after the second toss (verify). The phenomenon that $\boldsymbol{t}$ cannot be observed after the second toss clearly cannot take place in any probabilistic system.

Remark 1.4.1. In quantum mechanics, a state (1.10) is a pure state: it contains the maximal information of the system properties. The *mixed states of quantum systems* are discussed in Section 8.4.

Let us now continue the description of quantum systems by introducing two systems with basis states $|\boldsymbol{x}_1\rangle, \ldots, |\boldsymbol{x}_n\rangle$ and $|\boldsymbol{y}_1\rangle, \ldots, |\boldsymbol{y}_m\rangle$ respectively. As in the case of classical and probabilistic systems, the basis states of the compound system can be thought of as pairs $(|\boldsymbol{x}_i\rangle, |\boldsymbol{y}_j\rangle)$. It is natural to represent the general states of the compound system as

$$\sum_{i=1}^{n}\sum_{j=1}^{n} \alpha_{ij} |\boldsymbol{x}_i, \boldsymbol{y}_j\rangle, \tag{1.11}$$

where $\sum_{i=1}^{n}\sum_{j=1}^{n} |\alpha_{ij}|^2 = 1$. A natural mathematical framework for representations (1.11) is the *tensor product* of the state spaces of the subsystems, and now we will briefly explain what is meant by a tensor product.

Let H_n and H_m be two vector spaces with bases $\{\boldsymbol{x}_1, \ldots, \boldsymbol{x}_n\}$ and $\{\boldsymbol{y}_1, \ldots, \boldsymbol{y}_m\}$. The tensor product of spaces H_n and H_m is denoted by $H_n \otimes H_m$. Space $H_n \otimes H_m$ has ordered pairs $(\boldsymbol{x}_i, \boldsymbol{y}_j)$ as a basis, thus $H_n \otimes H_m$ has dimension mn. We also denote $(\boldsymbol{x}_i, \boldsymbol{y}_j) = \boldsymbol{x}_i \otimes \boldsymbol{y}_j$ and say that $\boldsymbol{x}_i \otimes \boldsymbol{y}_j$ is the tensor product of basis vectors $\boldsymbol{x}_i$ and $\boldsymbol{y}_j$. Tensor products of other vectors than basis vectors are defined by requiring that the product is bilinear:

$$\Big(\sum_{i=1}^{n} \alpha_i \boldsymbol{x}_i\Big) \otimes \Big(\sum_{j=1}^{m} \beta_j \boldsymbol{y}_j\Big) = \sum_{i=1}^{n}\sum_{j=1}^{m} \alpha_i \beta_j \boldsymbol{x}_i \otimes \boldsymbol{y}_j.$$

Since $\boldsymbol{x}_i \otimes \boldsymbol{y}_j$ form the basis of $H_n \otimes H_m$, the notion of the tensor product of vectors is perfectly well established, but notice carefully that the tensor product is not commutative. In connection with representing quantum systems, we usually omit the symbol $\otimes$ and use notations closer to the original idea of regarding $\boldsymbol{x}_i \otimes \boldsymbol{y}_j$ as a pair $(\boldsymbol{x}_i, \boldsymbol{y}_j)$:

$$|\boldsymbol{x}_i\rangle \otimes |\boldsymbol{y}_j\rangle = |\boldsymbol{x}_i\rangle |\boldsymbol{y}_j\rangle = |\boldsymbol{x}_i, \boldsymbol{y}_j\rangle .$$

If (1.11) can be represented as

$$\sum_{i=1}^{n}\sum_{j=1}^{m} \alpha_{ij} |\boldsymbol{x}_i\rangle |\boldsymbol{y}_j\rangle = \sum_{i=1}^{n}\sum_{j=1}^{m} \alpha_i \beta_j |\boldsymbol{x}_i\rangle |\boldsymbol{y}_j\rangle$$
$$= \Big(\sum_{i=1}^{n} \alpha_i |\boldsymbol{x}_i\rangle\Big)\Big(\sum_{j=1}^{m} \beta_j |\boldsymbol{y}_j\rangle\Big),$$

we say that the compound system is in *decomposable state.* Otherwise, the system state is *entangled.* It is plain to verify that the the notion "decomposable state" is independent of the bases chosen for each spaces.

Since $H_l \otimes (H_m \otimes H_n)$ is clearly isomorphic to $(H_l \otimes H_m) \otimes H_n$, we can omit the parentheses and refer to this tensor product as $H_l \otimes H_m \otimes H_n$. Thus, we can inductively define the tensor products of more than two spaces.

Example 1.4.3. A system of m quantum bits is described by an m-fold tensor product of two-dimensional Hilbert space H_2. Thus the system has basis states $|x_1\rangle |x_2\rangle \dots |x_m\rangle$, where $(x_1, \dots, x_m) \in \{0,1\}^m$. The dimension of this space is 2^m and the system of m qubits is referred to as the *quantum register of length m.*

We conclude this section by listing the key features of a finite-state quantum system in state-vector formalism.

- The basis states of an n-level quantum system form an orthonormal basis of the system state space, Hilbert space H_n. This basis can be chosen freely and it refers to a particular observable.
- A state of a quantum system is a unit-length vector

$$\alpha_1 |\boldsymbol{x}_1\rangle + \dots + \alpha_n |\boldsymbol{x}_n\rangle \,. \tag{1.12}$$

  These states are called superpositions of basis states, and they induce a probability distribution such that when one observes (1.12), a property $\boldsymbol{x}_i$ is seen with probability $|\alpha_i|^2$.
- If two quantum systems are depicted by using state spaces H_n and H_m, then the state space of the compound system consisting of these two subsystems is described by tensor product $H_m \otimes H_n$.
- The states of quantum systems change via unitary transformations.

2. Quantum Information

It must be first noted that, despite the title, *quantum information theory* is not the main topic of this chapter. Instead, we will merely concentrate on descriptional differences between presenting information by using classical and quantum systems. In fact, in this chapter, we will present the fundamentals of quantum information processing. A reader well aware of basic linear algebra will presumably have no difficulties in following this section, but for a reader feeling some uncertainty, we recommend consulting Sections 8.1, 8.2, and the initial part of Section 8.3. Moreover, the basic notions of linear algebra are outlined in Section 9.3.

2.1 Quantum Bits

Definition 2.1.1. *A quantum bit, qubit for short, is a two-level quantum system. Because there should not be any danger of confusion, we also say that the two-dimensional Hilbert space* H_2 *is a quantum bit. Space* H_2 *is equipped with a fixed basis* $B = \{|0\rangle, |1\rangle\}$*, a so-called computational basis. States* $|0\rangle$ *and* $|1\rangle$ *are also called basis states.*

A general *state* of a single quantum bit is a vector

$$c_0 |0\rangle + c_1 |1\rangle, \tag{2.1}$$

that has a unit length, i.e., $|c_0|^2 + |c_1|^2 = 1$. Numbers c_0 and c_1 are called *amplitudes* of $|0\rangle$ and $|1\rangle$, respectively.

We say that an *observation* of a quantum bit in state (2.1) will give 0 or 1 as an outcome with probabilities of $|c_0|^2$ and $|c_1|^2$, respectively.

Remark 2.1.1. The *coordinate representation* for quantum bits is chosen as $|0\rangle = \begin{pmatrix}1\\0\end{pmatrix}$, $|1\rangle = \begin{pmatrix}0\\1\end{pmatrix}$. The examples below concerning quantum gates assume that this representation is used.

Definition 2.1.2. *An operation on a qubit, called a unary quantum gate, is a unitary mapping* $U : H_2 \to H_2$.

In other words, a unary quantum gate defines a linear operation

$$|0\rangle \mapsto a\,|0\rangle + b\,|1\rangle \tag{2.2}$$
$$|1\rangle \mapsto c\,|0\rangle + d\,|1\rangle \tag{2.3}$$

such that the matrix

$$\begin{pmatrix} a & c \\ b & d \end{pmatrix}$$

is unitary, i.e.,

$$\begin{pmatrix} a & c \\ b & d \end{pmatrix}\begin{pmatrix} a^* & b^* \\ c^* & d^* \end{pmatrix} = \begin{pmatrix} 1 & 0 \\ 0 & 1 \end{pmatrix}.$$

Remark 2.1.2. In the coordinate representation, (2.2) can be written as

$$\begin{pmatrix} 1 \\ 0 \end{pmatrix} \mapsto a\begin{pmatrix} 1 \\ 0 \end{pmatrix} + b\begin{pmatrix} 0 \\ 1 \end{pmatrix} = \begin{pmatrix} a \\ b \end{pmatrix} = \begin{pmatrix} a & c \\ b & d \end{pmatrix}\begin{pmatrix} 1 \\ 0 \end{pmatrix},$$

and (2.3) can be written as

$$\begin{pmatrix} 0 \\ 1 \end{pmatrix} \mapsto c\begin{pmatrix} 1 \\ 0 \end{pmatrix} + d\begin{pmatrix} 0 \\ 1 \end{pmatrix} = \begin{pmatrix} c \\ d \end{pmatrix} = \begin{pmatrix} a & c \\ b & d \end{pmatrix}\begin{pmatrix} 0 \\ 1 \end{pmatrix}.$$

Notation a^* stands for the complex conjugate of complex number a. Notation A^* will be also used for different purposes, but this should cause no misunderstandings; the meaning of a particular *-symbol should be clear by the context. In what follows, notation $(a,b)^T$ is used to indicate the transposition, i.e.,

$$(a,b)^T = \begin{pmatrix} a \\ b \end{pmatrix}.$$

Example 2.1.1. Let us use the coordinate representation $|0\rangle = (1,0)^T$ and $|1\rangle = (0,1)^T$. Then the unitary matrix

$$M_\neg = \begin{pmatrix} 0 & 1 \\ 1 & 0 \end{pmatrix}$$

defines an action $M_\neg\,|0\rangle = |1\rangle$, $M_\neg\,|1\rangle = |0\rangle$. A unary quantum gate defined by $M_\neg$ is called a *quantum not-gate*.

Example 2.1.2. Let us examine a quantum gate defined by a unitary matrix

$$\sqrt{M_\neg} = \begin{pmatrix} \frac{1+i}{2} & \frac{1-i}{2} \\ \frac{1-i}{2} & \frac{1+i}{2} \end{pmatrix}.$$

The action of $\sqrt{M_\neg}$ is

$$\sqrt{M_\neg}\,|0\rangle = \frac{1+i}{2}\,|0\rangle + \frac{1-i}{2}\,|1\rangle\,, \tag{2.4}$$

$$\sqrt{M_\neg}\,|1\rangle = \frac{1-i}{2}\,|0\rangle + \frac{1+i}{2}\,|1\rangle\,. \tag{2.5}$$

Since

$$\left|\frac{1+i}{2}\right|^2 = \left|\frac{1-i}{2}\right|^2 = \frac{1}{2},$$

observation of (2.4) and (2.5) will give 0 or 1 as the outcome, both with a probability of $\frac{1}{2}$. Because $\sqrt{M_\neg}\sqrt{M_\neg} = M_\neg$, gate $\sqrt{M_\neg}$ is called the *square root of the not-gate.*

Example 2.1.3. Let us study a quantum gate W_2 defined by matrix

$$W_2 = \begin{pmatrix} \frac{1}{\sqrt{2}} & \frac{1}{\sqrt{2}} \\ \frac{1}{\sqrt{2}} & -\frac{1}{\sqrt{2}} \end{pmatrix}.$$

The action of W_2 is

$$W_2\,|0\rangle = \frac{1}{\sqrt{2}}\,|0\rangle + \frac{1}{\sqrt{2}}\,|1\rangle\,,$$

$$W_2\,|0\rangle = \frac{1}{\sqrt{2}}\,|0\rangle - \frac{1}{\sqrt{2}}\,|1\rangle\,.$$

Matrix W_2 is called a *Walsh matrix*, *Hadamard matrix*, or *Hadamard-Walsh matrix* and will eventually appear to be very useful. A very important feature of quantum gates that has been already mentioned implicitly is that they are *linear*, and therefore it suffices to know the action on the basis states. For example, if a Hadamard-Walsh gate acts on a state $\frac{1}{\sqrt{2}}(|0\rangle+|1\rangle)$, the outcome is

$$\begin{aligned} W_2(\frac{1}{\sqrt{2}}(|0\rangle + |1\rangle)) &= \frac{1}{\sqrt{2}}W_2\,|0\rangle + \frac{1}{\sqrt{2}}W_2\,|1\rangle \\ &= \frac{1}{\sqrt{2}}\frac{1}{\sqrt{2}}(|0\rangle + |1\rangle) + \frac{1}{\sqrt{2}}\frac{1}{\sqrt{2}}(|0\rangle - |1\rangle) \\ &= |0\rangle\,. \end{aligned}$$

Notice that the above equality reveals that $W_2W_2\,|0\rangle = |0\rangle$. Similarly, we can verify that $W_2W_2\,|1\rangle = |1\rangle$.

What happens to state $|0\rangle$ when Hadamard-Walsh gate W_2 is applied on it twice, is a very interesting matter. Figure 2.1 contains a scheme of that event.

The top row of the figure contains the initial state $|0\rangle$. When applied once, W_2 "splits" the state $|0\rangle$ into states $|0\rangle$ and $|1\rangle$; both will be present with amplitudes $\frac{1}{\sqrt{2}}$. This is depicted in the figure's middle row. The second application of W_2 "splits" $|0\rangle$ as before, but $|1\rangle$ is split slightly differently; $|1\rangle$

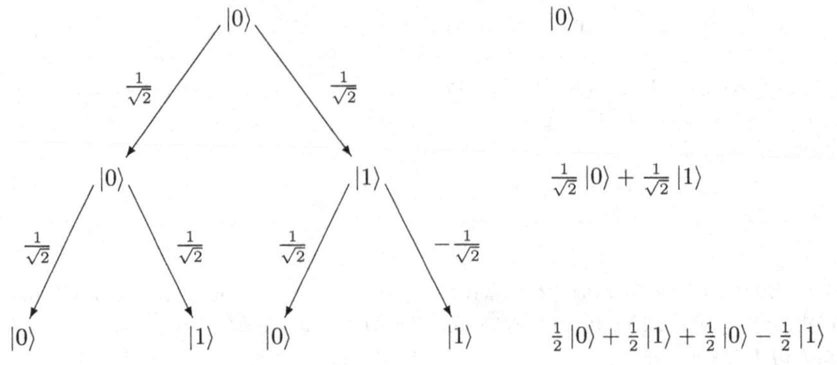

Fig. 2.1. Hadamard-Walsh-gate twice. The left-hand side depicts how the application of W_2 operates on states, whereas the corresponding states are written on the right side.

occurs with amplitude $-\frac{1}{\sqrt{2}}$. The bottom row of the figure describes the final state. Now, the amplitudes in the bottom row can be computed by following the path from top to bottom and multiplying all the amplitudes occuring in the path. For example, the amplitude of the left-most $|0\rangle$ in the bottom row is $\frac{1}{\sqrt{2}} \cdot \frac{1}{\sqrt{2}} = \frac{1}{2}$, whereas the amplitude of the right-most $|1\rangle$ is $\frac{1}{\sqrt{2}} \cdot (-\frac{1}{\sqrt{2}}) = -\frac{1}{2}$. When computing the outcome, we add up all the states in the lower right corner and get $|0\rangle$ as the outcome.

The effect that the amplitudes of states $|0\rangle$ sum to more than any of the summands is called *constructive interference.* On the other hand, the cancellation of states $|1\rangle$ is referred as to *destructive interference.*

Remark 2.1.3. The efficiency of quantum algorithms is based on the interference phenomenon. In the forthcoming chapters, we will learn about the most important quantum algorithms in more detail.

Example 2.1.4. Let F be defined as

$$F = \begin{pmatrix} 1 & 0 \\ 0 & -1 \end{pmatrix}.$$

Then, F acts as $F\,|0\rangle = |0\rangle$, $F\,|1\rangle = -\,|1\rangle$. Gate F is an example of unary quantum gates called *phase flips.* In general, phase flips are of the form

$$F_\theta = \begin{pmatrix} 1 & 0 \\ 0 & e^{i\theta} \end{pmatrix}$$

for a real θ (here we have denoted F_π by F).

2.2 Quantum Registers

A system of two quantum bits is a four-dimensional Hilbert space $H_4 = H_2 \otimes H_2$ having orthonormal basis $\{|0\rangle|0\rangle, |0\rangle|1\rangle, |1\rangle|0\rangle, |1\rangle|1\rangle\}$. We also write $|0\rangle|0\rangle = |00\rangle$, $|0\rangle|1\rangle = |01\rangle$, etc. A *state* of a two-qubit system is a unit-length vector

$$c_0|00\rangle + c_1|01\rangle + c_2|10\rangle + c_3|11\rangle, \tag{2.6}$$

so it is required that $|c_0|^2 + |c_1|^2 + |c_2|^2 + |c_3|^2 = 1$.

Observation of a two-qubit system in state (2.6) will give 00, 01, 10, and 11 as an outcome with probabilities of $|c_0|^2$, $|c_1|^2$, $|c_2|^2$, and $|c_3|^2$, respectively. On the other hand, if we choose to observe only one of the qubits, the standard rules of probabilities will apply. This means that an observation of the first (resp., second) qubit will give 0 or 1 with probabilities of $|c_0|^2 + |c_1|^2$ and $|c_2|^2 + |c_3|^2$ (resp., $|c_0|^2 + |c_2|^2$ and $|c_1|^2 + |c_3|^2$).

Remark 2.2.1. Notice that here the tensor product of vectors does not commute: $|0\rangle|1\rangle \neq |1\rangle|0\rangle$. We use linear ordering (write from left to right) to address the qubits individually.

Recall that the state $\boldsymbol{z} \in H_4$ of a two-qubit system is *decomposable* if $\boldsymbol{z}$ can be written as a product of states in H_2, $\boldsymbol{z} = \boldsymbol{x} \otimes \boldsymbol{y}$. A state that is not decomposable is *entangled.*

Example 2.2.1. State $\frac{1}{2}(|00\rangle + |01\rangle + |10\rangle + |11\rangle)$ is decomposable, since

$$\frac{1}{2}(|0\rangle|0\rangle + |0\rangle|1\rangle + |1\rangle|0\rangle + |1\rangle|1\rangle) = \frac{1}{\sqrt{2}}(|0\rangle + |1\rangle)\frac{1}{\sqrt{2}}(|0\rangle + |1\rangle),$$

as easily verified. On the other hand, state $\frac{1}{\sqrt{2}}(|00\rangle + |11\rangle)$ is entangled. To see this, assume on the contrary, that

$$\begin{aligned}\frac{1}{\sqrt{2}}(|00\rangle + |11\rangle) &= (a_0|0\rangle + a_1|1\rangle)(b_0|0\rangle + b_1|1\rangle)\\ &= a_0b_0|00\rangle + a_0b_1|01\rangle + a_1b_0|10\rangle + a_1b_1|11\rangle\end{aligned}$$

for some complex numbers a_0, a_1, b_0, and b_1. But then,

$$\begin{cases} a_0b_0 = \frac{1}{\sqrt{2}} \\ a_0b_1 = 0 \\ a_1b_0 = 0 \\ a_1b_1 = \frac{1}{\sqrt{2}}, \end{cases}$$

which is absurd.

Remark 2.2.2. If two qubits are in an entangled state $\frac{1}{\sqrt{2}}(|00\rangle + |11\rangle)$, then, observing one of them will give 0 or 1, both with a probability of $\frac{1}{2}$, but it is not possible to observe different values on the qubits. It is interesting to notice that the experiments have shown that this correlation can remain, even if the qubits are spatially separated by more than 10 km [86]. This distant correlation opens opportunities for quantum cryptography and quantum communication protocols. A pair of qubits in state $\frac{1}{\sqrt{2}}(|00\rangle + |11\rangle)$ is called an EPR pair.[1] Notice that if both qubits of the EPR pair are run through a Hadamard gate, the resulting state is again $\frac{1}{\sqrt{2}}(|00\rangle + |11\rangle)$, so it is impossible to give an ignorance interpretation for the EPR pair. EPR pairs are extremely important in many quantum protocols, and we will discuss their usefulness later.

Definition 2.2.1. *A binary quantum gate is a unitary mapping $H_4 \to H_4$. To define the operation of a binary quantum gate, we use coordinate representation $|00\rangle = (1,0,0,0)^T$, $|01\rangle = (0,1,0,0)^T$, $|10\rangle = (0,0,1,0)^T$, and $|11\rangle = (0,0,0,1)^T$.*

Example 2.2.2. Matrix

$$M_{\text{cnot}} = \begin{pmatrix} 1 & 0 & 0 & 0 \\ 0 & 1 & 0 & 0 \\ 0 & 0 & 0 & 1 \\ 0 & 0 & 1 & 0 \end{pmatrix}$$

defines a unitary mapping, whose action on the basis states is $M_{\text{cnot}}|00\rangle = |00\rangle$, $M_{\text{cnot}}|01\rangle = |01\rangle$, $M_{\text{cnot}}|10\rangle = |11\rangle$, $M_{\text{cnot}}|11\rangle = |10\rangle$. Gate M_{cnot} is called *controlled not*, since the second qubit *(target qubit)* is flipped if and only if the first *(control qubit)* is 1.

Other examples of multiqubit states and of their gates will be given after the following important definition.

Definition 2.2.2. *The tensor product, also called the Kronecker product, of $r \times s$ and $t \times u$ matrices*

$$A = \begin{pmatrix} a_{11} & a_{12} & \dots & a_{1s} \\ a_{21} & a_{22} & \dots & a_{2s} \\ \vdots & \vdots & \ddots & \vdots \\ a_{r1} & a_{r2} & \dots & a_{rs} \end{pmatrix} \quad \text{and } B = \begin{pmatrix} b_{11} & b_{12} & \dots & b_{1u} \\ b_{21} & b_{22} & \dots & b_{2u} \\ \vdots & \vdots & \ddots & \vdots \\ b_{t1} & b_{t2} & \dots & b_{tu} \end{pmatrix},$$

is an $rt \times su$ matrix defined as

[1] EPR stands for Einstein, Podolsky, and Rosen, who first regarded the distant correlation as the source of a paradox of quantum physics [88].

$$A \otimes B = \begin{pmatrix} a_{11}B & a_{12}B & \dots & a_{1s}B \\ a_{21}B & a_{22}B & \dots & a_{2s}B \\ \vdots & \vdots & \ddots & \vdots \\ a_{r1}B & a_{r2}B & \dots & a_{rs}B \end{pmatrix}.$$

If M_1 and M_2 are 2×2 matrices that describe unary quantum gates, then it is easy to verify that the joint actions of M_1 on the first qubit and M_2 on the second one are described by the matrix $M_1 \otimes M_2$. This also generalizes to quantum systems of any size: if matrices M_1 and M_2 define unitary mappings on Hilbert spaces H_n and H_m, then the $nm \times nm$ matrix $M_1 \otimes M_2$ defines a unitary mapping on space $H_n \otimes H_m$. The action of $M_1 \otimes M_2$ is exactly the same as the action of M_1 on H_n, followed by the action of M_2 on H_m (or vice versa). In particular, mapping $M_1 \otimes M_2$ cannot introduce entanglement between systems H_n and H_m.

Example 2.2.3. Using the Kronecker product of the previous definition, the coordinate representations $|0\rangle = \begin{pmatrix} 1 \\ 0 \end{pmatrix}$ and $|1\rangle = \begin{pmatrix} 0 \\ 1 \end{pmatrix}$ induce the coordinate representations for $|0\rangle|0\rangle = |0\rangle \otimes |0\rangle$, $|0\rangle|1\rangle$, $|1\rangle|0\rangle$, and $|1\rangle|1\rangle$ in a very natural way:

$$\begin{pmatrix} 1 \\ 0 \end{pmatrix} \otimes \begin{pmatrix} 1 \\ 0 \end{pmatrix} = \begin{pmatrix} 1 \\ 0 \\ 0 \\ 0 \end{pmatrix}, \quad \begin{pmatrix} 1 \\ 0 \end{pmatrix} \otimes \begin{pmatrix} 0 \\ 1 \end{pmatrix} = \begin{pmatrix} 0 \\ 1 \\ 0 \\ 0 \end{pmatrix},$$

$$\begin{pmatrix} 0 \\ 1 \end{pmatrix} \otimes \begin{pmatrix} 1 \\ 0 \end{pmatrix} = \begin{pmatrix} 0 \\ 0 \\ 1 \\ 0 \end{pmatrix}, \text{ and } \begin{pmatrix} 0 \\ 1 \end{pmatrix} \otimes \begin{pmatrix} 0 \\ 1 \end{pmatrix} = \begin{pmatrix} 0 \\ 0 \\ 0 \\ 1 \end{pmatrix}.$$

Similarly, we can get coordinate representations of triples of qubits, etc. Notice that the above coordinate representations agree with those of Definition 2.2.1 and that it is not a mere coincidence!

Example 2.2.4. Let $M_1 = M_2 = W_2$ be the Hadamard matrix of the previous section. Action on both qubits with a Hadamard-Walsh matrix can be seen as a binary quantum gate, whose matrix is defined by

$$W_4 = W_2 \otimes W_2 = \frac{1}{2} \begin{pmatrix} 1 & 1 & 1 & 1 \\ 1 & -1 & 1 & -1 \\ 1 & 1 & -1 & -1 \\ 1 & -1 & -1 & 1 \end{pmatrix},$$

and the action of W_4 can be written as

$$\begin{aligned} W_4 |x_0 x_1\rangle &= \frac{1}{2}\big(|0\rangle + (-1)^{x_0}|1\rangle\big)\big(|0\rangle + (-1)^{x_1}|1\rangle\big) \\ &= \frac{1}{2}\big(|00\rangle + (-1)^{x_1}|01\rangle + (-1)^{x_0}|10\rangle + (-1)^{x_0+x_1}|11\rangle\big) \quad (2.7) \end{aligned}$$

for any $x_0, x_1 \in \{0, 1\}$.

Remark 2.2.3. Note that state (2.7) is decomposable. This is, of course, natural since the initial state $|x_0 x_1\rangle$ is decomposable, and also W_4 decomposes into a product of unary gates, $W_4 = W_2 \otimes W_2$. On the other hand, M_{cnot} cannot be expressed as a tensor product of 2×2 matrices. This could, of course, be seen by assuming the contrary, as in Example 2.2.1, but we can use another argument. Consider a two-qubit system in an initial state $|00\rangle$. The action of the Hadamard-Walsh gate on the first qubit transforms this into a state

$$\frac{1}{\sqrt{2}}(|0\rangle + |1\rangle)\,|0\rangle = \frac{1}{\sqrt{2}}(|00\rangle + |10\rangle). \tag{2.8}$$

But the action of M_{cnot} turns the decomposable state (2.8) into an entangled state $\frac{1}{\sqrt{2}}(|00\rangle + |11\rangle)$. Because M_{cnot} introduces entanglement, it cannot be a tensor product of two unary quantum gates.

By a *quantum register* of the length m, we understand an ordered system of m qubits. The state space of such a system is the m-fold tensor product $H_{2^m} = H_2 \otimes \ldots \otimes H_2$, and the basis states are $\{|\boldsymbol{x}\rangle \mid \boldsymbol{x} \in \{0,1\}^m\}$. Identifying $\boldsymbol{x} = x_1 \ldots x_m$ with the binary representation of a number, we can also say that the basis states of an m-qubit register are $\{|a\rangle \mid a \in \{0, 1, \ldots, 2^m - 1\}\}$.

A peculiar feature of the exponential packing density associated with quantum systems can be plainly seen here: the general state of a system of m quantum bits is

$$c_0\,|0\rangle + c_1\,|1\rangle + \ldots + c_{2^m-1}\,|2^m - 1\rangle\,,$$

where $|c_0|^2 + |c_1|^2 + \ldots + |c_{2^m-1}|^2 = 1$. That is to say that a general description of an m two-state quantum system requires 2^m complex numbers. Hence, the description size of the system is exponential in its physical size.

The time evolution of an m-qubit system is determined by unitary mappings in H_{2^m}. The size of the matrix of such a mapping is $2^m \times 2^m$, also exponential in the physical size of the system.[2] A more detailed explanation of quantum register processing is provided in Section 3.2.3.

2.3 No-Cloning Theorem

So far, we have been talking about qubits, but everything also generalizes to *n-ary quantum digits*; If we have an alphabet $A = \{a_1, \ldots, a_n\}$, we can

[2] It should not be any more surprising that Feynman found the effective deterministic simulation of a quantum system difficult. Due to the interference effects, it also seems to be difficult to simulate a quantum system efficiently with a probabilistic computer.

identify the letters with basis states $|a_1\rangle$, ..., $|a_n\rangle$ of an n-level quantum system. We say that such a basis is a *quantum representation* of alphabet A. These key features of quantum representations are already listed at the end of the introductory chapter:

- A set of n elements can be identified with the vectors of an orthonormal basis of an n-dimensional complex vector space H_n. We call H_n a *state space.* When we have a fixed basis $|a_1\rangle$, ..., $|a_n\rangle$, we call these vectors *basis states.* Also, the basis that we choose to fix is usually called a *computational basis.*
- A general state of a quantum system is a unit-length vector in the state space. If $\alpha_1 |a_1\rangle + \ldots + \alpha_n |a_n\rangle$ is a state, then the system is seen in state a_i with a probability of $|\alpha_i|^2$.
- The state space of a compound system consisting of two subsystems is the tensor product of the subsystem state spaces.
- In this formalism, the state transformations are length-preserving linear mappings. It can be shown that these mappings are exactly the unitary mappings in the state space.

In the above list, we have operations that *can* be done with quantum systems (under this chosen formalism). We now present a somewhat surprising result called the "No-Cloning Theorem" due to W. K. Wootters and W. H. Zurek [91]. Consider a quantum system having n basis states $|a_1\rangle$, ..., $|a_n\rangle$. Let us denote the state space by H_n and specify that the state $|a_1\rangle$ is a "blank sheet state". A unitary mapping in $H_n \otimes H_n$ is called a *quantum copy machine,* if for any state $|\boldsymbol{x}\rangle \in H_n$,

$$U(|\boldsymbol{x}\rangle |a_1\rangle) = |\boldsymbol{x}\rangle |\boldsymbol{x}\rangle .$$

Theorem 2.3.1 (No-Cloning Theorem). *For $n > 1$, there is no quantum copymachine.*

Proof. Assume that a quantum copy machine exists, even if $n > 1$. Because $n > 1$, there are two orthogonal states $|a_1\rangle$ and $|a_2\rangle$. We should have $U(|a_1\rangle |a_1\rangle) = |a_1\rangle |a_1\rangle$ and $U(|a_2\rangle |a_1\rangle) = |a_2\rangle |a_2\rangle$, and also

$$U\Big(\frac{1}{\sqrt{2}}(|a_1\rangle + |a_2\rangle) |a_1\rangle\Big) = \Big(\frac{1}{\sqrt{2}}(|a_1\rangle + |a_2\rangle)\Big)\Big(\frac{1}{\sqrt{2}}(|a_1\rangle + |a_2\rangle)\Big)$$
$$= \frac{1}{2}(|a_1\rangle |a_1\rangle + |a_1\rangle |a_2\rangle + |a_2\rangle |a_1\rangle + |a_2\rangle |a_2\rangle).$$

But since U is linear,

$$U\Big(\frac{1}{\sqrt{2}}(|a_1\rangle + |a_2\rangle) |a_1\rangle\Big) = \frac{1}{\sqrt{2}}U(|a_1\rangle |a_1\rangle) + \frac{1}{\sqrt{2}}U(|a_2\rangle |a_1\rangle)$$
$$= \frac{1}{\sqrt{2}} |a_1\rangle |a_1\rangle + \frac{1}{\sqrt{2}} |a_2\rangle |a_2\rangle .$$

The above two representations for $U(\frac{1}{\sqrt{2}}(|a_1\rangle + |a_2\rangle) |a_1\rangle)$ do not coincide by the very definition of a tensor product, a contradiction. □

The No-Cloning Theorem thus states that there is no allowed operation (unitary mapping) that would produce a copy of an arbitrary quantum state. Notice also that in the above proof, we did not use unitarity; only the linearity of the time-evolution mapping was needed. If, however, we are satisfied with cloning only the *basis states*, there is a solution: let $I = \{1, \ldots, n\}$ be the set of indices. Partially defined mapping $f : I \times I \to I \times I$, $f(i, 1) = (i, i)$ is clearly injective, so we can complete the definition (in many ways) such that f becomes a permutation of $I \times I$ and still satisfies $f(i, 1) = (i, i)$. Let $f(i, j) = (i', j')$. Then the linear mapping defined by $U |a_i\rangle |a_j\rangle = |a_{i'}\rangle |a_{j'}\rangle$ is a permutation of basis vectors of $H_n \otimes H_n$, and any such permutation is unitary, as is easily verified. Moreover, $U(|a_i\rangle |a_1\rangle) = |a_i\rangle |a_i\rangle$, so U is a copy machine operation on basis vectors.

Remark 2.3.1. A. K. Pati and S. L. Braunstein introduced a principle complementary to the no-cloning principle [66]. We will return to this *no-deleting principle* in Section 8.4.6, where we also represent a stronger variant of the No-Cloning Theorem due to R. Jozsa.

Remark 2.3.2. Notice that, when defining the unitary mapping $U(|a_i\rangle |a_1\rangle) = |a_i\rangle |a_i\rangle$, we do not need to assume that the second system looks like the first one; the only thing we need is that the second system must have at least as many basis states as the first one. We could, therefore, also define a unitary mapping $U |a_i\rangle |b_1\rangle = |a_i\rangle |b_i\rangle$, where $|b_1\rangle, \ldots, |b_m\rangle$ $(m \geq n)$ are the basis states of some other system. What is interesting here is that we could regard the second system as a *measurement apparatus* designed to observe the state of the first system. Thus, $|b_1\rangle$ could be interpreted as the "initial pointer state", and U as a "measurement interaction". Measurements that can be described in this fashion, are called *von Neumann-Lüders measurements.* The measurement interaction is also discussed in Section 8.4.3.

2.4 Observation

So far, we have tacitly assumed that quantum systems are used for probabilistic information processing according to the following scheme:

- The system is first prepared in an initial basis state.
- Next, the unitary information processing is carried out.
- Finally, we observe the system to see the outcome.

What is missing in the above procedure is that we have not considered the possibility of making an intermediate observation[3] during unitary processing. In other words, we have not discussed the effect of observation upon the quantum system state. For a more systematic treatment of observation, the

[3] In physics literature, the term "observation" is usually replaced with "measurement".

reader is advised to study Section 8.3.2. For now, we will present, in a simplified way, the most widely used method for handling state changes during an observation procedure. Suppose that a system in state

$$\alpha_1 |\boldsymbol{x}_1\rangle + \ldots + \alpha_n |\boldsymbol{x}_n\rangle$$

is observed with $\boldsymbol{x}_k$ as the outcome. According to the *projection postulate*,[4] the state of the system after observation is $|\boldsymbol{x}_k\rangle$. Suppose, then, that we have a compound system in state

$$\sum_{i=1}^{n}\sum_{j=1}^{m} \alpha_{ij} |\boldsymbol{x}_i\rangle |\boldsymbol{y}_j\rangle \tag{2.9}$$

and the first system is observed with outcome $\boldsymbol{x}_k$ (notice that the probability of observing $\boldsymbol{x}_k$ is $P(k) = \sum_{j=1}^{m} |\alpha_{kj}|^2$). The projection postulate now implies that the postobservation state of the whole system is

$$\frac{1}{\sqrt{P(k)}} \sum_{j=1}^{m} \alpha_{kj} |\boldsymbol{x}_k\rangle |\boldsymbol{y}_j\rangle . \tag{2.10}$$

In other words, the initial state (2.9) of the system is projected to the subspace that corresponds to the observed state and renormalized to the unit length. It is now worth strongly emphasizing that the state evolution from (2.9) to (2.10) given by the projection postulate **is not consistent with the unitary time evolution**, since unitary evolution is always reversible; but there is no way to recover state (2.9) from (2.10). In fact, no explanation for the observation process which is consistent with quantum mechanics (using a certain interpretation) has ever been discovered. The difficulty arising when trying to find such an explanation is usually referred to as the *measurement paradox of quantum physics.*

However, instead of having intermediate observations that cause a "collapse of a state vector" from (2.9) to (2.10), we can have an auxiliary system with unitary measurement interaction (basis state copy machine) $|\boldsymbol{x}_i\rangle |\boldsymbol{x}_1\rangle \mapsto |\boldsymbol{x}_i\rangle |\boldsymbol{x}_i\rangle$. That is, we replace the collapse (2.9) $\mapsto$ (2.10) by a measurement interaction, which turns state

$$\Big(\sum_{i=1}^{n}\sum_{j=1}^{m} \alpha_{ij} |\boldsymbol{x}_i\rangle |\boldsymbol{y}_j\rangle\Big) |\boldsymbol{x}_1\rangle = \sum_{i=1}^{n}\sum_{j=1}^{m} \alpha_{ij} |\boldsymbol{x}_i\rangle |\boldsymbol{y}_j\rangle |\boldsymbol{x}_1\rangle \tag{2.11}$$

into

$$\sum_{i=1}^{n}\sum_{j=1}^{m} \alpha_{ij} |\boldsymbol{x}_i\rangle |\boldsymbol{y}_j\rangle |\boldsymbol{x}_i\rangle . \tag{2.12}$$

[4] The projection postulate can be seen as an ad hoc explanation for the state transform during the measurement process.

Even though the transformation from (2.11) to (2.12) appears different from (2.9) $\mapsto$ (2.10) at first glance, it has many similar features. In fact, using notation $P(k) = \sum_{j=1}^{m} |\alpha_{kj}|^2$ again, we can rewrite (2.12) as

$$\sum_{k=1}^{n} \Big(\frac{1}{\sqrt{P(k)}} \sum_{j=1}^{m} \alpha_{kj} |\boldsymbol{x}_k\rangle |\boldsymbol{y}_j\rangle \Big) \sqrt{P(k)} |\boldsymbol{x}_k\rangle . \tag{2.13}$$

Let us now interpret (2.12) (and (2.13)): the third system which we introduced shatters the whole system into n orthogonal subspaces, which are, for any $k \in \{1, \dots, n\}$ spanned by vectors $|\boldsymbol{x}_i\rangle |\boldsymbol{y}_j\rangle |\boldsymbol{x}_k\rangle$, $i \in \{1, \dots, n\}$, $j \in \{1, \dots, m\}$. The left multipliers of vectors $\sqrt{P(k)} |\boldsymbol{x}_k\rangle$ are unit-length vectors exactly the same as in (2.10). Moreover, the probability of observing $\boldsymbol{x}_k$ in the right-most system of (2.12) is $P(k)$. Therefore, it should be clear that, if quantum information processing continues independently of the third system, then the final probability distribution is the same if operation (2.9) $\mapsto$ (2.10) is replaced with operation (2.11) $\mapsto$ (2.12). For this reason, we will not consider intermediate observations in this book, but may refer to them only as notational simplifications of procedure (2.11) $\mapsto$ (2.12).

Anyway, it is an undenied fact that the observation procedure always disturbs the original system. If it becomes necessary to refer to a quantum system after observation, we will mainly adopt the projection postulate or even assume that the whole system is lost after observation.

2.5 Quantum Teleportation

In the cryptography community, there is a tradition to call two communicating parties not only by letters A and B but by names *Alice* and *Bob*. Since this convention brings some life to the text, we will also use it here.

Assume now that Alice has a single qubit in state

$$a|0\rangle + b|1\rangle , \tag{2.14}$$

but state (2.14) is unknown to Alice. Now Alice wants to send state (2.14) to Bob.

One, at least theoretical, possibility is to send Bob the whole two-state quantum system that is in state (2.14). We say that there is a *quantum channel* from Alice to Bob if this is possible. Similarly, if Alice can send classical bits to Bob, we say that there is a *classical channel* from Alice to Bob.

Now, we assume that there is no quantum channel from Alice to Bob, but a classical one exists. Alice's task, to send state (2.14) to Bob appears quite impossible. Notice that state (2.14) is unknown to Alice, so she cannot send Bob instructions for constructing (2.14).

Alice may try, for instance, to observe her state and then send Bob the outcome, 0 or 1. This attempt fails immediately if both a and b are nonzero; Bob cannot reconstruct state (2.14). Alice cannot make many observations either, because her observation always disturbs her qubit. Alice cannot make copies of her qubit for many observations, since copying of an unknown quantum state is impossible, as shown in Section 2.3.1. If an unlimited number of observations were allowed to Alice, she could get arbitrarily precise approximations of the *probabilities* of seeing 0 and 1 as outcomes, which is to say that she could get arbitrarily good *approximations* of numbers $|a|^2$ and $|b|^2$. But even the exact values of $|a|^2$ and $|b|^2$ are not enough for Bob to reconstruct state (2.14). This can be noticed immediately, for those values are identical for states

$$\frac{1}{\sqrt{2}}|0\rangle + \frac{1}{\sqrt{2}}|1\rangle \tag{2.15}$$

and

$$\frac{1}{\sqrt{2}}|0\rangle - \frac{1}{\sqrt{2}}|1\rangle . \tag{2.16}$$

Nevertheless, states (2.15) and (2.16) can behave quite differently, as seen by applying a Hadamard-Walsh gate (see Section 2.2) on both states: the outcomes will be $|0\rangle$ and $|1\rangle$, respectively (See Exercise 1).

In fact, it is impossible for Alice to send her quantum bit to Bob by only using a classical channel. To see this, it is enough to notice that classical information can be perfectly cloned: if there were a way to reconstruct state (2.14) from some classical information (which state (2.14) itself determines), then we would be able to make an unlimited number of reconstructions of (2.14). But this means that we would be able to create a quantum copy machine, and that was already proven impossible.

On the other hand, if Alice and Bob initially share an EPR pair, there is a way to execute the required task. This protocol, introduced in [11], is called *quantum teleportation.* [5]

We will now describe the teleportation protocol. The basic assumption is that Alice and Bob have two qubits in EPR state

$$\frac{1}{\sqrt{2}}|00\rangle + \frac{1}{\sqrt{2}}|11\rangle . \tag{2.17}$$

Notations are chosen such that the qubit on the left-hand side belongs to Alice and the right-hand side qubit belongs to Bob. In addition to qubits (2.17), Alice has her qubit to be teleported in state

[5] One could argue that to create an EPR pair, Alice and Bob should have a quantum channel. On the other hand, Alice and Bob could have a supply of EPR pairs generated when they last met.

$$a|0\rangle + b|1\rangle.$$

The compound state of all of the qubits will be denoted as

$$\begin{aligned}
&(a|0\rangle + b|1\rangle)\frac{1}{\sqrt{2}}(|00\rangle + |11\rangle) \\
&= \frac{a}{\sqrt{2}}|0\rangle|00\rangle + \frac{a}{\sqrt{2}}|0\rangle|11\rangle + \frac{b}{\sqrt{2}}|1\rangle|00\rangle + \frac{b}{\sqrt{2}}|1\rangle|11\rangle \\
&= \frac{a}{\sqrt{2}}|000\rangle + \frac{a}{\sqrt{2}}|011\rangle + \frac{b}{\sqrt{2}}|100\rangle + \frac{b}{\sqrt{2}}|111\rangle. \qquad (2.18)
\end{aligned}$$

Recall that only the right-most qubit belongs to Bob and that Alice has full access to the two qubits on the left.

Teleportation protocol

1. Alice performs the controlled not-operation on her qubit, using the left-most one as the control qubit (see Example 2.2.2). State (2.18) becomes then

$$\frac{a}{\sqrt{2}}|000\rangle + \frac{a}{\sqrt{2}}|011\rangle + \frac{b}{\sqrt{2}}|110\rangle + \frac{b}{\sqrt{2}}|101\rangle. \qquad (2.19)$$

2. Alice's next action is to make a Hadamard-Walsh transformation (recall Example 2.1.3) on the left-most qubit. The result is

$$\begin{aligned}
&\frac{a}{\sqrt{2}}\frac{1}{\sqrt{2}}(|0\rangle + |1\rangle)|00\rangle + \frac{a}{\sqrt{2}}\frac{1}{\sqrt{2}}(|0\rangle + |1\rangle)|11\rangle \\
&+ \frac{b}{\sqrt{2}}\frac{1}{\sqrt{2}}(|0\rangle - |1\rangle)|10\rangle + \frac{b}{\sqrt{2}}\frac{1}{\sqrt{2}}(|0\rangle - |1\rangle)|01\rangle,
\end{aligned}$$

which can be written as

$$\begin{aligned}
&\frac{a}{2}|000\rangle + \frac{a}{2}|100\rangle + \frac{a}{2}|011\rangle + \frac{a}{2}|111\rangle \\
&+ \frac{b}{2}|010\rangle - \frac{b}{2}|110\rangle + \frac{b}{2}|001\rangle - \frac{b}{2}|101\rangle. \qquad (2.20)
\end{aligned}$$

Taking apart Bob's qubit, state (2.20) can clearly be rewritten as

$$\begin{aligned}
&\frac{1}{2}|00\rangle a|0\rangle + \frac{1}{2}|10\rangle a|0\rangle + \frac{1}{2}|01\rangle a|1\rangle + \frac{1}{2}|11\rangle a|1\rangle \\
&+ \frac{1}{2}|01\rangle b|0\rangle - \frac{1}{2}|11\rangle b|0\rangle + \frac{1}{2}|00\rangle b|1\rangle - \frac{1}{2}|10\rangle b|1\rangle
\end{aligned}$$

and, moreover, as

$$\begin{aligned}
&\frac{1}{2}|00\rangle(a|0\rangle + b|1\rangle) + \frac{1}{2}|01\rangle(a|1\rangle + b|0\rangle) \\
&+ \frac{1}{2}|10\rangle(a|0\rangle - b|1\rangle) + \frac{1}{2}|11\rangle(a|1\rangle - b|0\rangle). \qquad (2.21)
\end{aligned}$$

3. Now Alice observes her two qubits. As the outcome, she sees 00, 01, 10, and 11, each with a probability of $\frac{1}{4}$. For brevity, we will refer to the projection postulate and summarize the resulting state (depending on the outcome) in the following table.

Alice's observation	Postobservation state
00	$\|00\rangle (a\|0\rangle + b\|1\rangle)$
01	$\|01\rangle (a\|1\rangle + b\|0\rangle)$
10	$\|10\rangle (a\|0\rangle - b\|1\rangle)$
11	$\|11\rangle (a\|1\rangle - b\|0\rangle)$

Recall now that the two left-most qubits are Alice's, whereas the right-most one belongs to Bob.
4. Alice sends Bob her observation result (two classical bits).
5. Bob makes the following:
   - If Alice's bits are 00, Bob makes nothing. His qubit is already in state $a|0\rangle + b|1\rangle$.
   - If Alice sent 01, Bob performs the not-operation on his qubit, thus obtaining the desired state.
   - If Alice sent 10, Bob performs the phase-flip operation (See Example 2.1.4) on his bit. Again, the resulting state is $a|0\rangle + b|1\rangle$.
   - If Alice's bits were 11, Bob first performs the not-operation and after that the phase flip.

Remark 2.5.1. It should be emphasized here that in the quantum teleportation described above, no physical qubit is transmitted, just the state of the qubit. It is also worth noticing that the quantum state is transmitted, not copied; the original state held by Alice is destroyed in the protocol.

Remark 2.5.2. The first experimental realizations of teleportation were reported in [15] and [16]. Shortly after, [39] reported an extended version of those first ones.

2.6 Superdense Coding

Superdense coding, introduced in [12], is the complementary action to quantum teleportation. Initially, Alice and Bob share an EPR pair, and there is a *quantum channel* from Alice to Bob. Now Alice wants to send *classical bits* to Bob. Superdense coding is a protocol where Alice sends one quantum bit to Bob, but the amount of transmitted information is *two* classical bits.

In theory, superdense coding is achieved as follows: Alice has two classical bits, b_1 and b_2, and Alice shares an EPR pair

$$\frac{1}{\sqrt{2}}|00\rangle + \frac{1}{\sqrt{2}}|11\rangle \tag{2.22}$$

with Bob. Again we assume that the qubit marked on the left is Alices, and the other one is Bob's.

Superdense coding protocol

1. If $b_1 = 1$, then Alice performs the phase flip on her qubit. If also $b_2 = 1$, then she also performs the not-operation on her qubit. What becomes of state (2.22) is summarized in the following table.

b_1 b_2	State after Alice's operations
0 0	$\frac{1}{\sqrt{2}}\|00\rangle + \frac{1}{\sqrt{2}}\|11\rangle$
0 1	$\frac{1}{\sqrt{2}}\|10\rangle + \frac{1}{\sqrt{2}}\|01\rangle$
1 0	$\frac{1}{\sqrt{2}}\|00\rangle - \frac{1}{\sqrt{2}}\|11\rangle$
1 1	$\frac{1}{\sqrt{2}}\|10\rangle - \frac{1}{\sqrt{2}}\|01\rangle$

2. Alice sends her qubit to Bob.
3. Now that Bob has access to both qubits, he runs them both through a two-qubit gate B defined by the matrix

$$B = \begin{pmatrix} \frac{1}{\sqrt{2}} & 0 & 0 & \frac{1}{\sqrt{2}} \\ 0 & \frac{1}{\sqrt{2}} & \frac{1}{\sqrt{2}} & 0 \\ \frac{1}{\sqrt{2}} & 0 & 0 & -\frac{1}{\sqrt{2}} \\ 0 & -\frac{1}{\sqrt{2}} & \frac{1}{\sqrt{2}} & 0 \end{pmatrix} \tag{2.23}$$

(verify that matrix B is unitary). It is then easy to verify, by direct calculation, that the following table (extending the previous one) is correct:

b_1 b_2	State after Alice's operations	State after Bob's operation
0 0	$\frac{1}{\sqrt{2}}\|00\rangle + \frac{1}{\sqrt{2}}\|11\rangle$	$\|00\rangle$
0 1	$\frac{1}{\sqrt{2}}\|10\rangle + \frac{1}{\sqrt{2}}\|01\rangle$	$\|01\rangle$
1 0	$\frac{1}{\sqrt{2}}\|00\rangle - \frac{1}{\sqrt{2}}\|11\rangle$	$\|10\rangle$
1 1	$\frac{1}{\sqrt{2}}\|10\rangle - \frac{1}{\sqrt{2}}\|01\rangle$	$\|11\rangle$

4. Bob observes his qubits. The table above shows that the bits b_1 and b_2 are recovered faithfully.

Remark 2.6.1. Regarding Alice's action in the beginning, it is sufficient and necessary to force the qubits into orthonormal states (recall from the beginning of Section of 2.2 that states $|00\rangle$, $|01\rangle$, $|10\rangle$, and $|11\rangle$ form an orthonormal set). Whenever this can be done, there always exists a unitary mapping that transforms these orthonormal states to basis states.

2.7 Exercises

1. Compute W_2W_2. Conclude that W_2 applied to state (2.15) (resp., 2.16) yields $|0\rangle$ (resp., $|1\rangle$)
2. Verify that matrix (2.23) is unitary.

3. Devices for Computation

To study computational processes, we have to fix a computational device first. In this chapter, we study Turing machines and circuits as models of computation. We use the standard notations of formal language theory and represent these notations briefly now. An *alphabet* is any set A. The elements of an alphabet A are called *letters.* The *concatenation* of sets A and B is a set AB consisting of strings formed of any element of A followed by any element of B. Especially, A^k is the set of strings of length k over A. These strings are also called *words.* The concatenation $w_1 w_2$ of words w_1 and w_2 is just the word w_1 followed by w_2. The *length* of word w is denoted by $|w|$ or $\ell(w)$ and defined as the number of the letters that constitute w. We also define A^0 as the set that contains only the *empty word* ϵ that has no letters, and $A^* = A^0 \cup A^1 \cup A^2 \cup \ldots$ is the set of all words over A. Mathematically speaking, A^* is the free monoid generated by the elements of A, having the concatenation as the monoid operation and ϵ as the unit element.

The attribute "uniform" is usually reserved for those models of computation, where the length of the input is not fixed but the input can be arbitrarily long. For instance, Turing machines can be regarded as uniform models of computation. On the other hand, circuits (which are handled in Section 3.2), are not uniform, since circuits must be constructed for each input size.

3.1 Uniform Computation

3.1.1 Turing Machines

The very classical model to describe computational processes is the *Turing machine*, or TM for short. To give a description of a Turing machine, we fix a finite set of basic information units called alphabet A, a finite set of *internal control states* Q and a *transition function*

$$\delta : Q \times A \to Q \times A \times \{-1, 0, 1\}. \tag{3.1}$$

Definition 3.1.1. *A (deterministic) Turing machine M over alphabet A is a sixtuple $(Q, A, \delta, q_0, q_a, q_r)$, where q_0, q_a, $q_r \in Q$ are the initial, accepting, and rejecting states respectively and δ as above.*

The reason for calling the Turing machine of the above definition *deterministic* is due to the form of the transition δ, which describes the dynamics (the computation) of the machine. Later we will introduce some other forms of Turing machines, but for now, we will give an interpretation for the above definition, mainly using the notations of [64]. As a *configuration* of a Turing machine we understand a triplet (q_1, x, y), where $q_1 \in Q$ and $x, y \in A^*$. For a configuration $c = (q_1, x, y)$, we say that the Turing machine is in state q_1 and that the *tape* contains word xy. If word y also begins with letter a_1, we say that the machine is *scanning* letter a_1 or *reading* letter a_1. To describe the dynamics determined by the transition function δ, we write $x = w_1 a$ and $y = a_1 w_2$, where $a, a_1 \in A$ and $w_1, w_2 \in A^*$. Thus, c can be written as $c = (q_1, w_1 a, a_1 w_2)$, and the transition δ defines also a transition rule from one configuration to another in a very natural way: if $\delta(q_1, a_1) = (q_2, a_2, d)$, then a configuration

$$c = (q_1, w_1 a, a_1 w_2)$$

can be transformed to

$$c' = (q_2, w_1, a a_2 w_2), \quad c' = (q_2, w_1 a, a_2 w_2), \quad \text{or} \quad c' = (q_2, w_1 a a_2, w_2),$$

depending on if $d = -1$, $d = 0$, or $d = 1$ respectively. This means that, under transition $\delta(q_1, a_1) = (q_2, a_2, d)$, the symbol a_1 being scanned is replaced with a_2 (we say that the machine *prints* a_2 to replace a_1); the machine enters state q_2, and begins to scan the symbol to the left of the previously scanned symbol, continues scanning the same location, or scans the symbol to the right of the previously scanned symbol, depending on $d \in \{-1, 0, 1\}$. We also say that the *Turing machine's read-write head* moves to the left, remains stationary, or moves to the right.

If δ defines a transition from c to c', we write $c \vdash c'$ and say that c' is a *successor* of c, c *yields* c', and that c *is followed by* c'. The transition from c to c' is called a *computational step*. The anomalous cases $c = (q_1, w, \epsilon)$ and $c = (q_1, \epsilon, w)$ (recall that ϵ was defined to be the empty word having no letters at all) are treated by introducing a *blank symbol* $\sqcup$, extending the definition of δ and replacing (q_1, w, ϵ) (resp. (q_1, ϵ, w)) with $(q_1, w, \sqcup)$, (resp. $(q_1, \sqcup, w)$) if necessary.

A *computation* of a Turing machine with input $w \in A^*$ is defined as a sequence of configurations $c_0, c_1, c_2, \ldots$ such that $c_0 = (q_0, \epsilon, w)$ and $c_i \vdash c_{i+1}$ for each i. We say that the computation *halts* if some c_i has no successors, or if the state symbol of the configuration of c_i is either q_a or q_r. In the former case, the computation is *accepting*, and in the latter case it is *rejecting*.

If a computation of a Turing machine T beginning with configuration (q_0, ϵ, w) leads into a halting configuration (q, w_1, w_2) in t computational steps, we say that T computes $w_1 w_2$ from the *input word* w in *time* t. Thus, a Turing machine can be seen as a device computing (partial) functions $A^* \to A^*$; but, we can also ignore the output $w_1 w_2$ and just say that the Turing machine T *accepts* the input w if it halts in an accepting state, and that T *rejects* the input w if it either does not halt or halts in a rejecting state. Thus, a Turing machine can also be seen as an *acceptor* that classifies the words into those which are accepted and those which are rejected.

Definition 3.1.2. *A set S of words over A is a recursively enumerable language if there is a Turing machine T such that T accepts w if and only if $w \in S$.*

Definition 3.1.3. *A set S is a recursive language if there is a Turing machine T such that each computation of T halts, and T accepts w if and only if $w \in S$.*

Families of recursively enumerable and recursive languages are denoted by **RE** and **R** respectively. By definition, $\mathbf{R} \subseteq \mathbf{RE}$, but it is a well-known fact that also $\mathbf{R} \neq \mathbf{RE}$; see [64], for instance.

It is a widespread belief that Turing machines capture the notion of algorithmic computability: whatever is algorithmically computable is also computable by a Turing machine. This strong belief should actually be called a hypothesis, which is known as *the Church-Turing thesis.* In this book we establish the notion of algorithm by using Turing machines. Turing machines provide a clear notion for computational resources like the time and the space used during computation. Moreover, by using Turing machines, it is easy to introduce the notions of probabilistic, nondeterministic and quantum computation. On the other hand, it is worth arguing that, for almost all practical purposes, the notion of the Turing machine is clumsy: even to describe very simple algorithms by using Turing machines requires a lengthy and fairly non-intuitive description of the transition rules. For the above reasons, we will represent the fundamental concepts of computability by using Turing machines, but we will also use more sophisticated notions of algorithms including Boolean circuits, quantum circuits and even pseudo-programming languages.

The above definitions of recursive and recursively enumerable languages represent two classes of algorithmic *decision problems.* When problem is a decision problem, it means that we are given a word w as an input and we should decide if w belongs to some particular language or not. It is clear that, for recursive languages, when the Turing machine accepts the particular language, it offers an algorithm for solving this problem. However, for recursively enumerable languages the corresponding Turing machine does not provide a good algorithmic solution. The reason for this is that even though the machine halts when the input w belongs to S, we do not know how long the computation lasts on individual input words w. Thus, we do not know, how long we should wait until we know that the machine will not halt, i.e., the decision $w \notin S$ cannot be made. We call all decision problems inside **R** *algorithmically solvable, recursively solvable,* or *decidable.* The decision problems outside **R** are called *recursively unsolvable* or *undecidable.*

The classification into recursively solvable and unsolvable problems is nowadays far too rough. A typical way to introduce some refinement is to consider the computational time required to solve a particular problem. Therefore, we now define some basic concepts connected to measuring the computation time. In complexity theory, we are usually interested in measuring the computation time only up to some multiplicative constant, and for that purpose, the following notations are useful:

Let f and g be functions $\mathbb{N} \to \mathbb{N}$. We write $f = O(g)$, if there are constants $c > 0$ and $n_0 \in \mathbb{N}$ such that $f(n) \leq cg(n)$, whenever $n \geq n_0$. If $f(n) \geq cg(n)$ whenever $n \geq n_0$, we write $f = \Omega(g)$. If $f = O(g)$ and $f = \Omega(g)$, then we write $f = \Theta(g)$ and say that f and g are of the same order.

Let M be a Turing machine that halts on each input and $T(n)$ be the maximum computation time on inputs having length n. We say that $T(n)$ is the *time complexity function* of M. A Turing machine M is a *polynomial-time Turing machine* if its time complexity function satisfies $T(n) = O(n^k)$ for some constant k. The family of decision problems that can be solved by polynomial-time Turing machines is denoted by **P**. Using widespread jargon, we say that the decision problems in **P** are *tractable*, while the other ones are *intractable*.

3.1.2 Probabilistic Turing Machines

We can modify the definition of a deterministic Turing machine to provide a model for probabilistic computation. The necessary modification is to replace the transition function with *transition probability distribution*

$$\delta : Q \times A \times Q \times A \times \{-1, 0, 1\} \to [0, 1].$$

Value $\delta(q_1, a_1, q_2, a_2, d)$ is interpreted as the probability that, when the machine in state q_1 reads symbol a_1, it will print a_2, enter the state q_2, and move the head to direction $d \in \{-1, 0, 1\}$.

Definition 3.1.4. *A probabilistic Turing machine M over alphabet A is a sixtuple $(Q, A, \delta, q_0, q_a, q_r)$, where q_0, q_a, $q_r \in Q$ are the initial, accepting, and rejecting states respectively. It is required that for all $(q_1, a_1) \in Q \times A$*

$$\sum_{(q_2, a_2, d) \in Q \times A \times \{-1,0,1\}} \delta(q_1, a_1, q_2, a_2, d) = 1.$$

From this time on, to avoid fundamental difficulties in the notion of computability, we will agree that all the values of $\delta(q_1, a_1, q_2, a_2, d)$ are *rational.*[1]

The computation of a probabilistic Turing machine is not so straightforward a concept as the deterministic computation. We say that the configuration $c = (q_1, w_1 a, a_1 w_2)$ *yields* (or *is followed by*) any configuration

[1] This agreement can, of course, be criticized, but the feasibility of a machine working with arbitrary real number probabilities is also questionable.

$c' = (q_2, w_1, aa_2w_2)$ (resp. $c' = (q_2, w_1a, a_2w_2)$ and $c' = (q_2, w_1aa_2, w_2)$) with probability $p = \delta(q_1, a_1, q_2, a_2, 1)$ (resp. with probability $p = \delta(q_1, a_1, q_2, a_2, 0)$ and $p = \delta(q_1, a_1, q_2, a_2, -1)$). If c yields c' with probability p, we write $c \vdash_p c'$. Let $c_0, c_1, \ldots, c_t$ be a sequence of configurations such that $c_i \vdash_{p_{i+1}} c_{i+1}$ for each i. Then we say that c_t *is computed from* c_0 in t steps with probability $p_1p_2\cdots p_t$. If $p_1p_2\cdots p_t \neq 0$, we also say that $c_0 \vdash_{p_1} c_1 \vdash_{p_2} \ldots \vdash_{p_t} c_t$ is a *computation* of a probabilistic Turing machine.

Remark 3.1.1. A reader who thinks that the notion of probabilistic computation based on probabilistic Turing machines does not correspond very well to the idea of an algorithm utilizing random bits is perfectly right! In many ways a simpler model for probabilistic computations could have been given by using *nondeterministic Turing machines.* The reason for presenting this model is to make the traditional definition of a quantum Turing machine (see e.g. [13]) clearer.

Unlike in a deterministic computation, a single configuration can now yield several different configurations with probabilities that sum up to 1. Thus, the total computation of a probabilistic Turing machine can be seen as a tree having configurations as nodes. The initial configuration is the root and each node c has configurations followed by c as descendants. Thus, a computation of the probabilistic machine in question is just a single branch in the total computation tree. All computations can be expressed by the terms of the probabilistic systems we considered in the introductory chapter: assume that taking t steps, a probabilistic Turing machine starting from an initial configuration computes configurations $c_1, \ldots, c_m$ with probabilities $p_1, \ldots, p_m$ such that $p_1 + \ldots + p_m = 1$. We can then say that the *total configuration* of a probabilistic machine at time t is a probability distribution

$$p_1[c_1] + \ldots + p_m[c_m] \tag{3.2}$$

over the *basis configurations* $c_1, \ldots, c_m$. As in the introductory chapter, we can define a vector space, which we now call the *configuration space*, that has all of the potential basis configurations as the basis vectors, and a general configuration (3.2) is a vector in the configuration space having non-negative coordinates that sum up to 1. On the other hand, there is now an essentially more complicated feature compared to the introductory chapter: there is a countable infinity of basis vectors, so our configuration space is countably infinite-dimensional.

Regarding this latter representation of total configurations of a probabilistic Turing machine, the reader may already guess how to introduce the notion of *quantum Turing machines*, but we will still continue with probabilistic computations and study some *acceptance models*.

When talking about time-bounded computation in connection with probabilistic Turing machines, we usually assume that all the computations have the same length. In other words, we assume that all the computations are synchronized so well that they reach a halting configuration at the same time, so that all the branches of the computation tree have the same length. We can thus regard a probabilistic Turing machine as a facility for computing the probability distribution of outcomes. For instance, if the purpose of a particular probabilistic machine is to solve a decision problem, then some of the computations may end up in an accepting state, but some may also end up in a rejecting state. What then do we mean by saying that a probabilistic machine accepts a string or a language? In fact, there are several different choices, some of which are given in the following definitions:

- Class **NP** is the family of languages S that can be accepted in polynomial time by some probabilistic Turing machine M in the following sense: a word w is in the language S if and only if M accepts w with nonzero probability (see Remark 3.1.2).
- Class **RP** is the family of languages S that can be accepted in polynomial time with some probabilistic Turing machine M in the following way: if $w \in S$, then M accepts w with a probability of at least $\frac{1}{2}$, but if $w \notin S$, then M always rejects w.
- Class **coRP** is the class of languages consisting exactly of the complements of those in **RP**. Notice that **coRP** is not the complement of **RP** among all the languages.
- We define **ZPP** = **RP** $\cap$ **coRP** .
- Class **BPP** is the family of languages S that are accepted by a probabilistic Turing machine M such that if $w \in S$, then M accepts with a probability of at least $\frac{2}{3}$, and if $w \notin S$, then M rejects with a probability of at least $\frac{2}{3}$.

Remark 3.1.2. The definition of the class **NP** (standing for *nondeterministic polynomial time*) given here is not the usual one. A much more traditional definition would be given by using the notion of a *nondeterministic Turing machine*, which can be obtained from the probabilistic ones by ignoring the probabilities. In other words, a nondeterministic Turing machine has a *transition relation* $\delta \subseteq Q\times A\times Q\times A\times\{-1,0,1\}$, which tells whether it is *possible* for a configuration c to yield another configuration c'. More precisely, the fact that $(q_1, a_1, q_2, a_2, d) \in \delta$ means that, if the machine is in state q_1 scanning symbol a_1, then it is possible to replace a_1 with a_2, move the head to direction d, and enter state q_2. However, this model resembles the probabilistic computation very closely: indeed, the notion of "computing c' from c in t steps with nonzero probability" is replaced with the notion that "there is a possibility to compute c' form c in t steps", which does not seem to make any difference. The acceptance model for nondeterministic Turing machines also looks like the one that we defined for probabilistic **NP**-machines. A word w is accepted if and only if it is possible to reach an accepting final configuration in polynomial time.

It can also be argued that the class **NP** does not correspond very well to our intuition of practical computation. For example, if each configuration yields two distinct ones each with a probability of $\frac{1}{2}$ and the computation lasts t steps, there are 2^t final configurations, each computed from the initial one with a probability of $\frac{1}{2^t}$. However, we say that the machine accepts a word if and only if at least one of these final configurations is accepting, but it may happen that only one final configuration is accepting, and we cannot distinguish the acceptance probabilities $\frac{1}{2^t}$ (accepting) and 0 (rejecting) practically without running the machine $\Omega(2^t)$ times. But this would usually make the computation last *exponential time* since, if the machine reads the whole input, then $t \geq n$, where n is the length of the input. By its very definition, each deterministic Turing machine is also a probabilistic one (always working with a probability of 1), and therefore $\mathbf{P} \subseteq \mathbf{NP}$. However, it is a long-standing open problem in theoretical computer science whether $\mathbf{P} \neq \mathbf{NP}$.

Remark 3.1.3. Class **RP** (*randomized polynomial time*) corresponds more closely to the notion of realizable effective computation. If $w \notin S$, then this fact is revealed by any computation. On the other hand, if $w \in S$ we learn this with a probability of at least $\frac{1}{2}$. This means that an **RP**-Turing machine has no false positives, but it may give a false negative, however, with a probability of less than $\frac{1}{2}$. By repeating the whole computation k times, the probability of making false decision $w \notin S$ can thus be reduced to at most $\frac{1}{2^k}$. A probabilistic Turing machine with **RP**-style acceptance is also called *Monte Carlo Turing machine.*

Remark 3.1.4. If a language S belongs to **ZPP**, then there are Monte Carlo Turing machines M_1 and M_2 accepting S and the complement of S respectively. By combining these two machines, we obtain an algorithm that can be repeatedly used to make the correct decision with certainty. For if M_1 gives answer $w \in S$ one time, it surely means that $w \in S$ since M_1 has no false positives. If M_2 gives an answer $w \in A^* \setminus S$, then we know that $w \notin S$ since M_2 has no false positives either. In both cases, the probability of a false answer is at most $\frac{1}{2}$, so by repeating the procedure of running both machines k times, we obtain with a probability of at least $1 - \frac{1}{2^k}$ the certainty that either $w \in S$ or $w \notin S$. Thus, we can say that a **ZPP**-algorithm works like a deterministic algorithm whose *expected running time* is polynomial. Notation **ZPP** stands for *Zero error Probability in Polynomial time.* A **ZPP**-algorithm is called a *Las Vegas-algorithm.*

Remark 3.1.5. The definition of **BPP** merely contains the idea that we accept languages or, to put it in other words, that we solve decision problems using a probabilistic Turing machine that is required to give a correct answer with a probability that is larger than $\frac{2}{3}$. Notation **BPP** stands for *Bounded error Probability in Polynomial time.* The constant $\frac{2}{3}$ is arbitrary and any $c \in (\frac{1}{2}, 1)$ would define the same class. This is because we can efficiently increase the success probability by defining a Turing machine that runs the same computation several times, and then taking the majority of results as the answer. It is also widely believed that the class **BPP** is the class of problems that are *efficiently solvable.* This belief is known as the *Extended Church-Turing* thesis.

Earlier in this section it was mentioned that, according to the Church-Turing thesis, Turing machines are capable of capturing the notion of computability. However, it seems that probabilistic Turing machines with any of the above acceptance models also fit very well into the notion of "algorithmic computability". Is it true that any language which is accepted by a probabilistic Turing machine, regardless of the acceptance mode, can also be accepted by an ordinary Turing machine? The answer is yes, and it can be justified as follows: since the probabilities were required to be rational, we can always simulate the computation tree of a probabilistic Turing machine with an ordinary one that sequentially performs all possible computations and the probabilities associated to them. In other words, knowledge of the probabilities on how a probabilistic algorithm works also allows us to compute the probability distribution deterministically.[2]

Therefore, the notion of probabilistic computation does not shatter the border between decidability and undecidability. However, any deterministic Turing machine can also be seen as a probabilistic machine having only one choice for each transition. Therefore, it is clear that

$$\mathbf{P} \subseteq \mathbf{RP}, \quad \mathbf{P} \subseteq \mathbf{ZPP}, \text{ and } \mathbf{P} \subseteq \mathbf{BPP}. \tag{3.3}$$

The simulation of a probabilistic machine by a deterministic one is done by sequentially computing all the possible computations of a probabilistic machine. Since the number of different computations may be exponential (recall Remark 3.1.2), the simulation may also take exponential time. Hence, we can ask whether probabilistic computation is more *efficient* than deterministic computation, i.e., if some of the inclusions (3.3) are strict. Or is it true that there is a cunning way to imitate deterministically a probabilistic computation such that the simulation time would not grow exponentially? The answer to such questions are unknown so far.

3.1.3 Multitape Turing Machines

A straightforward generalization of a Turing machine is a *multitape Turing machine.* This model of computation becomes apparent when speaking about space-bounded computation.

---

[2] The restriction to rational probabilities plays an essential role here. This restriction can be criticized, but from the practical point of view, rational probabilities should be sufficient. Some authors allow all "computable numbers" as probabilities, but this would introduce more problems, such as: Which numbers are computable? Which is the computational model to "compute these numbers"? Can the comparisons of "computable numbers" (needed to decide, e.g., to decide whether the acceptance probability is at least $\frac{2}{3}$) be made efficiently?

Definition 3.1.5. *A (deterministic) Turing machine with k tapes over an alphabet A is a sixtuple $(Q, A, \delta, q_0, q_a, q_r)$, where q_0, q_a, $q_r \in Q$ are the initial, accepting, and rejecting states respectively, and $\delta : Q \times A^k \to Q \times (A \times \{-1, 0, 1\})^k$ is the transition function.*

A configuration of a k-tape Turing machine is a $2k + 1$-tuple

$$c = (q, x_1, y_1, x_2, y_2, \dots, x_k, y_k), \tag{3.4}$$

where $q \in Q$ and $x_i, y_i \in A^*$. For a configuration (3.4) we say that q is the *current state*, and that $x_i y_i$ is the *content of the ith tape.* Let $x_i = v_i a_i$ and $y_i = b_i w_i$. Then we say that b_i is the *currently scanned symbol on the ith tape.* If

$$\delta(q, b_1, b_2, \dots, b_k) = (r, (c_1, d_1), (c_2, d_2), \dots, (c_k, d_k)),$$

where $r \in Q$, $c_i \in A$, and $d_i \in \{-1, 0, 1\}$, then the configuration c yields $c' = (r, x'_1, y'_1, x'_2, y'_2, \dots, x'_k, y'_k)$, where $(x'_i, y'_i) = (v_i, a_i c_i w_i)$, if $d_i = -1$, $(x'_i, y'_i) = (v_i a_i, c_i w_i)$, if $d_i = 0$, and $(x'_i, y'_i) = (v_i a_i c_i, w_i)$, if $d_i = 1$.

The concepts associated with computing, such as acceptation, rejection, computational time, etc., are defined for multitape Turing machines exactly in the same fashion as for ordinary Turing machines. But when talking about the space consumed by the computation, it may seem fair to ignore the space occupied by the input and the space needed for writing the output, and merely talk about the space consumed by the computation. By using multitape machines, this is achieved as follows: we choose $k > 2$ and identify the first tape as an *input tape* and the last tape as an *output tape.* The machine must never change the contents of the input tape, and the head position in the output tape can never move to the left. If, using such a model for computation, one reaches a final configuration $c = (q, u_1, v_1, u_2, v_2, \dots, u_k, v_k)$, then the space required during the computation is defined to be $\sum_{i=2}^{k-1} |u_i v_i|$.

Again, we may ask whether the computational capacity of multitape Turing machines is greater than that of ordinary ones. This time, we can even guarantee that the computation capacity of multitape machines is very close to the capacity of a single-tape machine: it is left as an exercise to show that whatever is computable by a multitape machine in time t is also computable by a single-tape Turing machine in time $O(t^2)$.

3.1.4 Quantum Turing Machines

To provide a model for quantum computation, one can consider a straightforward generalization of a probabilistic Turing machine, replacing the probabilities with transition amplitudes. Thus, there should be a *transition amplitude function*

$$\delta : Q \times A \times Q \times A \times \{-1, 0, 1\} \to \mathbb{C}$$

such that $\delta(q_1, a_1, q_2, a_2, d)$ gives the *amplitude* that whenever the machine is in state q_1 scanning symbol a_1, it will replace a_1 with a_2, enter state q_2, and move the head to direction $d \in \{-1, 0, 1\}$.

Definition 3.1.6. *A quantum Turing machine (QTM) over alphabet A is a sixtuple $(Q, A, \delta, q_0, q_a, q_r)$, where q_0, q_a, $q_r \in Q$ are the initial, accepting, and rejecting states. The transition amplitude function must satisfy*

$$\sum_{(q_2,a_2,d)\in Q\times A\times\{-1,0,1\}} |\delta(q_1, a_1, q_2, a_2, d)|^2 = 1$$

for any $(q_1, a_1) \in Q \times A$.

Notions such as "configuration yields another" are straightforward generalizations of those associated with probabilistic computation. We can also generalize the acceptance models **NP**, **RP**, **ZPP**, and **BPP** to correspond to quantum computation. For any family **F** of languages accepted by some classical model of computation (deterministic or probabilistic), we usually write **QF** to stand for the corresponding class in quantum computation. However, there is already quite a well-established tradition to denote the quantum counterpart of **BPP** [3] not by **QBPP**, but by **BQP**, and the quantum counterpart of **P**,[4] is usually denoted by **EQP** to stand for *Exact acceptation by a Quantum computer in Polynomial time.* The quantum counterpart of **NP** is usually denoted by **NQP**.

As in probabilistic computation, a general configuration of a quantum Turing machine can be seen as a combination

$$\alpha_1 |c_1\rangle + \ldots + \alpha_m |c_m\rangle \tag{3.5}$$

of basis configurations. Taking the basis configurations as an orthonormal basis of an infinite-dimensional vector space, we see that the general configurations, *superpositions*, are merely the unit-length vectors of the configuration space. Moreover, the transition amplitude function determines a linear mapping M_δ in the state space. From the theory of quantum mechanics (to be precise: from the formalism that we are using) there now arises a further requirement on the transition amplitude function δ: *the linear mapping M_δ in the configuration space determined by δ should be* **unitary**.

It turns out that the unitarity of the mapping M_δ can be determined by the local conditions on the transition amplitude function (see [13], [45], and [63]), but the conditions are quite technical, and we will not represent them here. Instead, we will study *quantum circuits* as a model for quantum

[3] The quantum counterpart of **BPP** is the family of languages accepted by a polynomial-time quantum Turing machine with a correctness probability of at least $\frac{2}{3}$.

[4] The family of languages accepted by a polynomial-time quantum Turing machine with a probability of 1.

computation in the next chapter. The reason for this choice is that, compared to QTMs, a quantum circuit model is much more straightforward to use for describing quantum algorithms. In fact, all of the quantum algorithms that are studied in chapters 4-6 are given by using quantum circuit formalism.

We end this section by examining a couple of properties of quantum Turing machines. The first thing to consider is that the transition of a QTM determines a unitary and, therefore, a reversible time evolution in the configuration space. However, ordinary Turing machines can be irreversible, too.[5] The question is how powerful the reversible computation is. Can we design a reversible Turing machine that performs each particular computational task? A positive answer to this question was first given by Lecerf in [53], who intended to demonstrate that *the Post Correspondence Problem*[6] remains undecidable even for injective morphisms. Lecerf's constructions were later extended by Ruohonen [78], but Bennett in [8] was the first to give a model for reversible computation simulating the original, possibly irreversible, computation with *constant slowdown* but possibly with a huge increase in the space consumed (see also [46]).

Bennett's work was at least partially motivated by a thermodynamic problem: according to Landauer [52], an irreversible overwriting of a bit causes at least $kT \ln 2$ joules of energy dissipation.[7] This theoretical lower bound can always be ignored by using reversible computation, which is possible according to the results of Lecerf and Bennett.

Bennett's construction of a reversible Turing machine uses a three-tape Turing machine with *input tape*, *history tape*, and *output tape*. Reversibility is obtained by simulating the original machine on the input tape, thereby writing down *the history* of the computation, i.e., the transition rules that have been used so far, onto the history tape. When the machine stops, the output is copied from the input tape to the empty output tape, and the computation is run backward (also a reversible procedure) to erase the history tape for future use. The amount of space this construction consumes is proportional to the computation time of the original computation, but by applying the erasure of the history tape recursively, the space requirement can be reduced even to $O(s(n) \log t(n))$ at the same time using time $O(t(n) \log t(n))$, where s and t are the original space and time consumption [9]. For space/time trade-offs for reversible computing, see also [54].

---

[5] We call an ordinary Turing machine *reversible* if each configuration admits a unique precessor.

[6] The Post Correspondence Problem, or PCP for short, was among the first computational problems that were shown to be undecidable. The undecidability of the PCP was established by Emil Post in [71]. The importance of the PCP lies in the simple combinatorial formulation of the problem; the PCP is very useful for establishing other undecidability results and studying the boundary between decidability and undecidability.

[7] Here $k = 1.380658 \cdot 10^{-23}$J/K is *Bolzmann's constant*, and T is the absolute temperature.

Thus, it is established that whatever is computable by a Turing machine is also computable by a reversible Turing machine, which can be seen as a quantum Turing machine as well. Therefore, QTMs are at least as powerful as ordinary Turing machines. The next question which arises is whether QTMs can do something that the ordinary Turing machines cannot. The answer to this question is no; the reason is the same as for probabilistic Turing machines: we can use an ordinary Turing machine to simulate the computation of a QTM by remembering all the coexisting configurations in a superposition and their amplitudes. Arguing like this, we have to assume that all the amplitudes can be written in a way which can be described finitely. For example, it may be required that all the transition amplitudes can be written as $x + yi$, where x and y are rational. This assumption can, of course, be criticized, but the feasibility of a model having arbitrary complex number amplitudes is also questionable. It is, however, true that, on some operations on quantum systems, nonrational amplitudes, such as $\frac{1}{\sqrt{2}}$, would be very natural. Therefore, in practice, we will also use amplitudes $x+yi$ where x and y are not rational, but only if there is an "efficient" way to find rational approximations for x and y. It is left to the reader to verify that in each example in this book, the amplitudes not expressible by rational numbers can be well approximated by amplitudes which can be expressed by using rational numbers.

We have good reasons to believe (as did Feynman) that QTMs cannot be *efficiently* simulated by ordinary ones, not even by probabilistic ones. One major reason to believe this lies in Shor's factoring algorithm [81]: there is a polynomial-time, probabilistic quantum algorithm for factoring integers, but any classical polynomial-time algorithm for that purpose has not been found, not even a probabilistic one, despite the huge effort.

The next and final issue connected to QTMs in this section is the existence of a *universal quantum Turing machine.* We say that a Turing machine is *universal* if it can simulate the computation of any other Turing machine in the following sense: the description (the transition rules) and the input of an arbitrary Turing machine M are given to the universal machine U suitably encoded, and the universal machine U performs the computation of M on the encoded string. We could also say that the universal Turing machine U is *programmable* so as to perform any Turing machine computation. It can be shown that a universal Turing machine with 5 states having alphabet size 5 exists; see [74]. In [31], D. Deutsch proved the existence of a *universal quantum Turing machine* capable of simulating any other QTM *with arbitrary precision*, but Deutsch did not consider the efficiency of the simulation. Bernstein and Vazirani [13] supplied a construction of a universal QTM, which simulates any other Turing machine such that, given the description of a QTM M, a natural number T, and $\epsilon > 0$, the universal machine can simulate M with precision ϵ for T steps in time polynomial in T and $\frac{1}{\epsilon}$.

3.2 Circuits

3.2.1 Boolean Circuits

Recall that a Turing machine can be regarded as a facility for computing a partially defined function $f : A^* \to A^*$. We will now fix $A = \mathbb{F}_2 = \{0, 1\}$ to be the *binary alphabet*[8] and consider *the Boolean circuits* computing functions $\{0,1\}^n \to \{0,1\}^m$. As the basic elements of these circuits we choose functions $\wedge : \mathbb{F}_2^2 \to \mathbb{F}_2$ (the *logical and-gate*) defined by $\wedge(x_1, x_2) = 1$ if and only if $x_1 = x_2 = 1$, $\vee : \mathbb{F}_2^2 \to \mathbb{F}_2$ (the *logical or-gate*) defined as $\vee(x_1, x_2) = 0$ if and only if $x_1 = x_2 = 0$ and the *logical not-gate* $\mathbb{F}_2 \to \mathbb{F}_2$ defined by $\neg x = 1 - x$.[9]

A Boolean circuit is an acyclic, directed graph whose nodes are labelled either with *input variables*, *output variables*, or *logical gates* $\wedge$, $\vee$, and $\neg$. An *input variable node* has no incoming arrows, while an *output variable node* has no outcoming arrows but exactly one incoming arrow. Nodes $\wedge$, $\vee$, and $\neg$ have 2, 2, and 1 incoming arrows respectively. In connection with Boolean circuits, the arrows of the graph are also called *wires*. The number of the nodes of a circuit is called the *complexity* of the Boolean circuit.

A Boolean circuit with n input variables $x_1, \ldots, x_n$ and m output variables $y_1, \ldots, y_m$ naturally defines a function $\mathbb{F}_2^n \to \mathbb{F}_2^m$: the input is encoded by giving the input variables values 0 and 1, and each gate computes a primitive function $\wedge$, $\vee$ or $\neg$. The value of the function is given in the sequence of the output variables $y_1, \ldots, y_m$.

Example 3.2.1. The Boolean circuit in Figure 3.1 computes the function $f : \mathbb{F}_2^2 \to \mathbb{F}_2^2$ defined by $f(0,0) = (0,0)$, $f(0,1) = (0,1)$, $f(1,0) = (0,1)$, and $f(1,1) = (1,0)$. Thus, $f(x_1, x_2) = (y_1, y_2)$, where y_1 is $x_1 + x_2$ modulo 2, and y_2 is the *carry bit.*

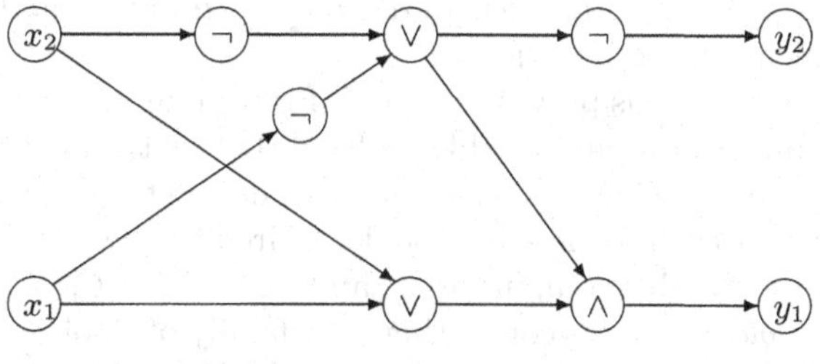

Fig. 3.1. A Boolean circuit computing the sum of bits x_1 and x_2

[8] $\mathbb{F}_2$ stands for the *binary field* with two elements 0 and 1, the addition and the multiplication defined by $0+0 = 1+1 = 0$, $0+1 = 1$, $0 \cdot 0 = 0 \cdot 1 = 0$, $1 \cdot 1 = 1$. Thus $-1 = 1$ and $1^{-1} = 1$ in $\mathbb{F}_2$.

[9] Notations $x_1 \wedge x_2$ and $x_1 \vee x_2$ are also used instead of $\wedge(x_1, x_2)$ and $\vee(x_1, x_2)$.

Emil Post has proven a powerful theorem characterizing the *complete sets of truth functions* [70] which implies that *all* functions $\mathbb{F}_2^n \to \mathbb{F}_2^m$ can be computed by using Boolean circuits with gates $\wedge$, $\vee$, and $\neg$. A set S of gates is called *universal* if all functions $\mathbb{F}_2^n \to \mathbb{F}_2$ can be constructed by using the gates in S. In what follows, we will demonstrate that $S = \{\wedge, \vee, \neg\}$ is a universal set of gates. The fact that all functions $\mathbb{F}_2^n \to \mathbb{F}_2^m$ are also computable follows from the observation that one can design any function $f : \mathbb{F}_2^n \to \mathbb{F}_2^m$ using m functions which each compute a single bit.

First, it is easy to verify that $\vee(x_1, \vee(x_2, x_3))$ equals to 0 if and only if $x_1 = x_2 = x_3 = 0$. For short, we denote $\vee(x_1, \vee(x_2, x_3)) = x_1 \vee x_2 \vee x_3$. Clearly this generalizes to any number of variables: using only function $\vee$ on two variables, it is possible to construct a function $x_1 \vee x_2 \vee \ldots \vee x_n$, which takes value 0 if and only if all the variables are 0. Similarly, using only $\wedge$ on two variables, one can construct function $x_1 \wedge x_2 \wedge \ldots \wedge x_n$, which takes value 1 if and only if all the variables are 1. We define a function $M_{\boldsymbol{a}}$ for each $\boldsymbol{a} = (a_1, \ldots, a_n) \in \mathbb{F}_2^n$ by

$$M_{\boldsymbol{a}}(x_1, \ldots, x_n) = \phi_1(x_1) \wedge \phi_2(x_2) \wedge \ldots \wedge \phi_n(x_n),$$

where $\phi_i(x_i) = \neg x_i$, if $a_i = 0$ and $\phi_i(x_i) = x_i$, if $a_i = 1$. Thus, functions $M_{\boldsymbol{a}}$ can be constructed by using only $\wedge$ and $\neg$. Moreover, $M_{\boldsymbol{a}}(x_1, \ldots, x_n) = 1$ if and only if each $\phi_i(x_i) = 1$ for each i, but

$$\phi_i(x_i) = \begin{cases} x_i, & \text{if } a_i = 1 \\ \neg x_i, & \text{if } a_i = 0, \end{cases}$$

so $\phi_i(x_i) = 1$ if and only if $x_i = a_i$. It follows that $M_{\boldsymbol{a}}$ is the *characteristic function* of singleton $\{\boldsymbol{a}\}$: $M_{\boldsymbol{a}}(\boldsymbol{x}) = 1$ if and only if $\boldsymbol{x} = \boldsymbol{a}$. Using the characteristic functions $M_{\boldsymbol{a}}$ it is easy to build any function f: $f = M_{\boldsymbol{a}_1} \vee M_{\boldsymbol{a}_2} \vee \ldots \vee M_{\boldsymbol{a}_k}$, where $\boldsymbol{a}_1$, $\boldsymbol{a}_2$, ..., $\boldsymbol{a}_k$ are exactly the elements of $\mathbb{F}_2^n$ for which f takes a value of 1.

Boolean circuits thus provide a facility for computing functions $\mathbb{F}_2^n \to \mathbb{F}_2^m$, where the number of input variables is fixed. Let us then consider a function $f : \{0,1\}^* \to \{0,1\}$ defined on any binary string. Let f_n be the restriction of f on $\{0,1\}^n$. For each n there is a Boolean circuit C_n with n input variables and one output variable computing f_n, and we say that C_0, C_1, C_2, C_3, ..., is a *family of Boolean circuits* computing f. A family of Boolean circuits having only one output node can be also used to recognize languages over the binary alphabet: we regard a binary string with length n as *accepted* if the output of C_n is 1, otherwise that string is *rejected.*

So far we have demonstrated that, for an arbitrary function $f_n : \mathbb{F}_2^n \to \mathbb{F}_2$ there is a circuit computing f_n. Thus, for an arbitrary binary language L there is a family C_0, C_1, C_2, ..., accepting L. Does this violate Church-Turing's thesis? There are only numerably many Turing machines, so there are only numerably many binary languages that can be accepted by Turing machines. But all binary languages (there are non-numerably many such languages) can be accepted by Boolean circuit families. Is the notion of a circuit family a stronger algorithmic device than a Turing machine? The answer is that the notion of a circuit family is not an algorithmic device at all. The reason is that the way in which we constructed the circuit C_n requires exact knowledge of the values of f_n. In other words, without knowing the values of f, we *do not know how to construct* the circuit family C_0, C_1, C_2, ...; we can only state that such a family *exists* but, without the construction, we cannot say that the circuit family C_0, C_1, C_2, ... is an algorithmic device.

We say that a language L has *uniformly polynomial circuits* if there exists a Turing machine M that on input 1^n (n consecutive 1's) outputs the graph of circuit C_n using space $O(\log n)$, and the family C_0, C_1, C_2, ... accepts L. For the proof of the following theorem, consult [64].

Theorem 3.2.1. *A language L has uniformly polynomial circuits if and only if $L \in \mathbf{P}$.*

Since our main concern in this book is polynomial-time quantum computation, we will use quantum circuit formalism which is analogous to the uniformly polynomial circuits.

3.2.2 Reversible Circuits

In Section 2 we introduced unary and binary quantum gates as unitary mappings. This is a generalization of the notion of *reversible* gates, so let us now study some properties of reversible gates. A reversible gate on m bits is a permutation on $\mathbb{F}_2^m$, so there are m inputs and output bits. Clearly there are $(2^m)!$ reversible gates on m bits.

Example 3.2.2. Function $T : \mathbb{F}_2^3 \to \mathbb{F}_2^3$, $T(x_1, x_2, x_3) = (x_1, x_2, x_1x_2 - x_3)$ defines a reversible gate on three bits. This gate is called the *Toffoli gate.* The Toffoli gate does not change bits x_1 and x_2, but computes the not-operation on x_3 if and only if $x_1 = x_2 = 1$. The symbol in Figure 3.2 is used to signify the Toffoli gate.

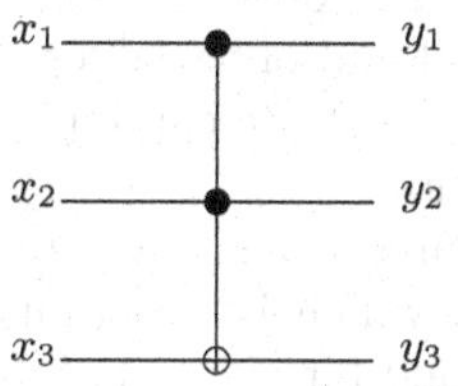

Fig. 3.2. The Toffoli gate

A *reversible circuit* is a permutation on $\mathbb{F}_2^n$ composed of reversible gates. Thus, we make no fundamental distinction between reversible gates and reversible circuits, but the only difference is contextual: we usually assume that there is a fixed set of reversible gates, and an arbitrarily large reversible circuit can be built from them.

We call a set R of reversible gates *universal* if any reversible circuit $C : \mathbb{F}_2^n \to \mathbb{F}_2^n$ can be constructed using the gates in R, constants, and some workspace. This means that, using the gates in R, we can construct a permutation $f : \mathbb{F}_2^{n+m} \to \mathbb{F}_2^{n+m}$ such that there is a fixed constant vector $(c_1, \ldots, c_m) \in \mathbb{F}_2^m$ such that

$$f(x_1, \ldots, x_n, c_1, \ldots, c_m) = (y_1, \ldots, y_n, d_1, \ldots, d_m),$$

where $(y_1, \ldots, y_n) = C(x_1, \ldots, x_n)$. Moreover, in this construction we also allow *bit permutations* that swap any two bits:

$$(\ldots, x_i, \ldots, x_j, \ldots) \mapsto (\ldots, x_j, \ldots, x_i, \ldots),$$

but these bit permutations could, of course, be avoided by requiring only that f simulates C in such a way that the constants are scattered among the variables in fixed places, and that the output can be read on fixed bit positions. If a universal set R consists of a single gate, this gate is said to be universal. When a universal set R of reversible gates is fixed, we again say that the *complexity* of a circuit C (with respect to the set R) is the number of the gates in C.

Earlier we demonstrated that a Boolean circuit constructed of irreversible gates $\wedge$, $\vee$, and $\neg$ can compute any function $\mathbb{F}_2^n \to \mathbb{F}_2^m$, which especially implies that, by using these gates, we can build any reversible circuit, as well. Recalling that any Turing machine can also be simulated by a reversible Turing machine, it is no longer surprising that all Boolean circuits can be simulated by using only reversible gates. In fact:

- Not-gates are already reversible, so we can use them directly.
- And-gates are irreversible, and therefore we simulate any such by using a Toffoli gate $T(x_1, x_2, x_3) = (x_1, x_2, x_1x_2 - x_3)$. Notice that $T(x_1, x_2, 0) = (x_1, x_2, x_1x_2)$, so, using the constant 0 on the third component, a Toffoli gate computes the logical and of variables x_1 and x_2.
- Since $x_1 \vee x_2 = \neg(\neg x_1 \wedge \neg x_2)$, we can replace all the $\vee$-gates in a Boolean circuit by $\wedge$- and $\neg$-gates.
- The fanout (multiple wires leaving a gate) is simulated by a controlled not-gate $C : \mathbb{F}_2^2 \to \mathbb{F}_2^2$, which is a permutation defined by $C(x_1, x_2) = (x_1, x_1 - x_2)$, so the second bit is negated if and only if the first bit is one. Again, $C(x_1, 0) = (x_1, x_1)$, so, by using the constant 0 we can duplicate the first bit x_1.

Because of the above properties, we can construct a reversible circuit for each Boolean circuit simulating the original one. This contruction uses not-gates,

Toffoli gates, controlled not-gates and the constant 0. But using the constant 1, the Toffoli gates can also be used to replace the not- and controlled not-gates: $T(1,1,x_1) = (1,1,\neg x_1)$ and $T(1,x_1,x_2) = (1,x_1,x_1 - x_2)$, so the Toffoli gates with constants 0 and 1 are sufficient to simulate any Boolean circuit with gates $\wedge$, $\vee$ and $\neg$. Since Boolean circuits can be used to build up an arbitrary function $\mathbb{F}_2^n \to \mathbb{F}_2^m$, we have obtained the following theorem.

Theorem 3.2.2. *A Toffoli gate is a universal reversible gate.*

Remark 3.2.1. A more straightforward proof for the above theorem was given by Toffoli in [87]. It is interesting to note that there are universal two-qubit gates for quantum computation, [33] but it can be shown that there are no universal reversible two-bit gates. In fact, there are only $4! = 24$ reversible two-bit gates, and all of them are *linear*, i.e., they can all be expressed as $T(\boldsymbol{x}) = A\boldsymbol{x} + \boldsymbol{b}$, where $\boldsymbol{b} \in \mathbb{F}_2^2$ and A is an invertible matrix over the binary field. Thus, any function composed of them is also linear, but there are also nonlinear reversible gates, such as the Toffoli gate.

3.2.3 Quantum Circuits

We identify again the bit strings $\boldsymbol{x} \in \mathbb{F}_2^m$ and an orthogonal basis $\{|\boldsymbol{x}\rangle \mid \boldsymbol{x} \in \mathbb{F}_2^m\}$ of a 2^m-dimensional Hilbert space H_{2^m}. To represent linear mappings $\mathbb{F}_2^m \to \mathbb{F}_2^m$, we adopt the coordinate representation $|\boldsymbol{x}\rangle = \boldsymbol{e}_i = (0,\ldots,1,\ldots,0)^T$, where $\boldsymbol{e}_i$ is a column vector having zeroes elsewhere but 1 in the ith position, if the components of $\boldsymbol{x} = (x_1,\ldots,x_m)$ form a binary representation of number $i-1$.

A reversible gate f on m bits is a permutation of $\mathbb{F}_2^m$, so any reversible gate also defines a linear mapping in H_{2^m}. There is a $2^m \times 2^m$ *permutation matrix* $M(f)$ [10] representing this mapping, $M(f)_{ij} = 1$, if $f(\boldsymbol{e}_j) = \boldsymbol{e}_i$, and $M(f)_{ij} = 0$ otherwise.

Example 3.2.3. In $\mathbb{F}_2^3$, we denote $|000\rangle = (1,0,\ldots,0)^T$, $|001\rangle = (0,1,\ldots,0)^T$, ..., $|111\rangle = (0,0,\ldots,1)^T$. The matrix representation of the Toffoli gate is

$$M(T) = \begin{pmatrix} I_2 & & & 0 \\ & I_2 & & \\ & & I_2 & \\ 0 & & & M_\neg \end{pmatrix},$$

where I_2 is the 2×2 identity matrix and $M_\neg$ is the matrix of the not-gate (Example 2.1.1).

Quantum gates generally are introduced as straightforward generalizations of reversible gates.

---

[10] A permutation matrix is a matrix having entries 0 and 1, exactly one 1 in each row and column.

Definition 3.2.1. *A quantum gate on m qubits is a unitary mapping in $H_2 \otimes \ldots \otimes H_2$ (m times), which operates on a fixed number (independent of m) of qubits.*

Because $M(f)^*_{ij} = 1$ if and only if $f(\boldsymbol{e}_i) = \boldsymbol{e}_j$, $M(f)^*$ represents the inverse permutation to f. Therefore, a permutation matrix is always unitary, and reversible gates are special cases of quantum gates. The notion of a *quantum circuit* is obtained from that of a reversible circuit by replacing the reversible gates by quantum gates. The only difference between quantum gates and quantum circuits here is again contextual: we require that a quantum circuit must be composed of quantum gates, such that each gate operates only on a bounded number of qubits (the bound is the same for each gate).

Definition 3.2.2. *A quantum circuit on m qubits is a unitary mapping on H_{2^m}, which can be represented as a concatenation of a finite set of quantum gates.*

Since reversible circuits are also quantum circuits, we have already discovered the fact that whatever is computable by a Boolean circuit is also computable by a quantum circuit. It is also interesting to compare the computational power of polynomial-size quantum circuits (that is, quantum circuits containing polynomially many quantum gates) and polynomial-time QTMs. A. Yao [92] has shown that their computational powers coincide.

It would be also interesting to know which kinds of gates are needed for quantum computing. The very first answer to that was given by David Deutsch, who demonstrated that there exists a three-qubit *universal gate* for quantum computing [32].[11]

It has turned out that the *controlled not*-gate (see Example 2.2.2) plays a most important role in quantum computing.

Theorem 3.2.3 (**[5]**). *All quantum circuits can be constructed by using only controlled not-gates and unary gates.*

Remark 3.2.2. Even though 2-qubit gates are enough for quantum computing, 2-bit gates are not enough for classical reversible computing. This is quite easy to see; recall Remark 3.2.1.

Remark 3.2.3. It is known that to implement quantum computing, only real numbers are needed [13] and that quantum computing can be regarded as "discrete" [13]. The theorem below gives even more information. Even though these important consequences of Shi's theorem [80] were known before, it expresses many things in a very compact form.

---

[11] By a universal gate, we mean here a quantum gate that can be used to *approximate* all quantum networks. Moreover, usually it is assumed that some *ancilla* qubits can be used. The values of those qubits are set previously in some fixed manner, and the values of those qubits are ignored when reading the result.

Theorem 3.2.4 ([80]). *All quantum circuits can be constructed (in the approximate sense) by using only Toffoli gates and Hadamard-Walsh gates.*

Remark 3.2.4. By the so-called Gottesman-Knill theorem [40], all quantum circuits consisting only of Hadamard-Walsh gates and of controlled not-gates, can be simulated with only polynomial loss of efficiency by using classical circuits.

In the forthcoming chapters, we will use mainly the quantum circuit formalism for representing quantum algorithms. To conclude this section, we mention a very important thorem of *Solovay and Kitaev.* [50]

Theorem 3.2.5. *Assume that S is a finite set of unary quantum gates that can approximate any unary quantum gate up to an arbitrary precision. There exists a constant C depending on S only and $c \approx 4$ such that any unary quantum gate can be approximated up to precision ϵ by using at most $C \log^c(\frac{1}{\epsilon})$ gates from set S.*

The above theorem, together with Theorem 3.2.3, implies that an n-gate quantum circuit can be simulated by using $O(n \log^c(\frac{n}{\epsilon}))$ gates from a universal set.

4. Fast Factorization

In this chapter we represent Shor's quantum algorithm for factoring integers. Shor's algorithm can be better understood after studying *quantum Fourier transforms*. The issues related to Fourier transforms and other mathematical details are handled in Chapter 9, but a reader having a solid mathematical knowledge of these concepts is advised to ignore the references to Chapter 9.

4.1 Quantum Fourier Transform

4.1.1 General Framework

Let $G = \{g_1, g_2, \ldots, g_n\}$ be an abelian group (we will use the additive notations) and $\{\chi_1, \chi_2, \ldots, \chi_n\}$ the characters of G (see Section 9.2). The functions $f : G \to \mathbb{C}$ form a complex vector space V, addition and scalar multiplication are defined pointwise. If f_1, $f_2 \in V$, then the *standard inner product* (see Section 9.3) of f_1 and f_2 is defined by

$$\langle f_1 \mid f_2 \rangle = \sum_{k=1}^{n} f_1^*(g_k) f_2(g_k).$$

Any inner product induces a *norm* by $||f|| = \sqrt{\langle f \mid f \rangle}$. In Section 9.2 it is demonstrated that the functions $B_i = \frac{1}{\sqrt{n}}\chi_i$ form an orthonormal basis of the vector space, so each $f \in V$ can be represented as

$$f = c_1 B_1 + c_2 B_2 + \ldots + c_n B_n,$$

where c_i are complex numbers called the *Fourier coefficients* of f. The *discrete Fourier transform* of $f \in V$ is another function $\widehat{f} \in V$ defined by $\widehat{f}(g_i) = c_i$. Since the functions B_i form an orthonormal basis, we see easily that $c_i = \langle B_i \mid f \rangle$, so

$$\widehat{f}(g_i) = \frac{1}{\sqrt{n}} \sum_{k=1}^{n} \chi_i^*(g_k) f(g_k). \tag{4.1}$$

The Fourier transform satisfies *Parseval's identity*

$$||\widehat{f}|| = ||f||, \tag{4.2}$$

which will be important in the sequel.

Let H be a finite quantum system capable of representing the elements of G. This means that $\{|g\rangle \mid g \in G\}$ is an orthonormal basis of some quantum system H. To obtain the matrix representations for linear mappings, we use the coordinate representation $|g_i\rangle = \boldsymbol{e}_i = (0, \ldots, 1, \ldots, 0)^T$ (all coordinates 0 except 1 in the ith position; see Section 9.3). The general states of the system are the unit-length linear combinations of the basis states. Thus, a general state

$$c_1 |g_1\rangle + c_2 |g_2\rangle + \ldots + c_n |g_n\rangle \tag{4.3}$$

of H can be seen as a mapping

$$f : G \to \mathbb{C}, \quad \text{where } f(g_i) = c_i \text{ and } ||f|| = 1, \tag{4.4}$$

and vice versa, each mapping (4.4) defines a state of H.

Definition 4.1.1. *The quantum Fourier transform (QFT) is the operation*

$$\sum_{i=1}^{n} f(g_i) |g_i\rangle \mapsto \sum_{i=1}^{n} \widehat{f}(g_i) |g_i\rangle . \tag{4.5}$$

In other words, QFT is just the ordinary Fourier transform of a function $f : G \to \mathbb{C}$ determined by the state (4.3) via (4.4). It is clear by the formula (4.1) that the QFT (4.5) is linear, but by Parseval's identity (4.2), QFT is even a unitary operation in H. As seen in Exercise 1, for basis vectors $|g_i\rangle$ the operation (4.5) becomes

$$|g_i\rangle \to \frac{1}{\sqrt{n}} \sum_{k=1}^{n} \chi_k^*(g_i) |g_k\rangle , \tag{4.6}$$

and hence it is clear that in basis $|g_i\rangle$, the matrix of the QFT is

$$\frac{1}{\sqrt{n}} \begin{pmatrix} \chi_1^*(g_1) & \chi_1^*(g_2) & \cdots & \chi_1^*(g_n) \\ \chi_2^*(g_1) & \chi_2^*(g_2) & \cdots & \chi_2^*(g_n) \\ \vdots & \vdots & \ddots & \vdots \\ \chi_n^*(g_1) & \chi_n^*(g_2) & \cdots & \chi_n^*(g_n) \end{pmatrix}. \tag{4.7}$$

What kind of quantum circuit is needed to implement (4.6)? The problem which arises is that, in a typical situation, $n = |G| = \dim(H)$ is large, but usually H is a tensor product of smaller spaces. However, quantum circuit operations were required to be local, not affecting a large number of quantum digits at the same time. In other words, the problem is how we can decompose the QFT matrix (4.7) into a tensor product of small matrices, or into a product of few matrices that can be expressed as a tensor product of matrices operating only on some small subspaces.

To approach this problem, let us assume that $G = U \oplus V$ is a direct sum of the subgroups U and V. Let $r = |U|$ and $s = |V|$, hence $|G| = rs$. Let also U and V be represented by some quantum systems H_U and H_V with orthonormal bases

$$\{|u_1\rangle, \ldots, |u_r\rangle\} \text{ and } \{|v_1\rangle, \ldots, |v_s\rangle\}$$

respectively. Then, the tensor product $H_U \otimes H_V$ represents G in a very natural way: each $g \in G$ can be uniquely expressed as $g = u + v$, where $u \in U$ and $v \in V$, so we represent $g = u + v$ by $|u\rangle|v\rangle$. Since we have a decomposition $\widehat{G} = \widehat{U} \times \widehat{V}$, all the characters of G can be written as

$$\chi_{ij}(g) = \chi_{ij}(u + v) = \chi_i^U(u)\chi_j^V(v),$$

where χ_i^U and χ_j^V are characters of U and V respectively (see Section 9.2.1), and

$$(i, j) \in \{1, \ldots, r\} \times \{1, \ldots, s\}.$$

Thus, the Fourier transforms can also be decomposed:

Lemma 4.1.1 (Fourier transform decomposition). *Let $G = U \oplus V$ be a direct product of subgroups U and V and $\{|u\rangle|v\rangle \mid u \in U, v \in V\}$ a quantum representation of the elements of G. Then*

$$|u_i\rangle|v_j\rangle \mapsto \Big(\frac{1}{\sqrt{r}}\sum_{k=1}^{r}(\chi_i^U(u_k))^*|u_k\rangle\Big)\Big(\frac{1}{\sqrt{s}}\sum_{l=1}^{s}(\chi_j^V(v_l))^*|v_l\rangle\Big) \tag{4.8}$$

$$= \frac{1}{\sqrt{rs}}\sum_{k=1}^{r}\sum_{l=1}^{s}\chi_{ij}^*(u_k + v_l)|u_k\rangle|v_l\rangle \tag{4.9}$$

is the Fourier transform on G.

The decomposition of the Fourier transform may be applied recursively to groups U and V. It is also interesting to notice that the state (4.9) is decomposable.

4.1.2 Hadamard-Walsh Transform

Let us study a special case $G = \mathbb{F}_2^m$. All the characters of the additive group of $\mathbb{F}_2^m$ are

$$\chi_{\boldsymbol{y}}(\boldsymbol{x}) = (-1)^{\boldsymbol{x}\cdot\boldsymbol{y}},$$

where $\boldsymbol{y} \in \mathbb{F}_2^m$. To implement the corresponding QFT we have to find a quantum circuit that performs the operation

$$|\boldsymbol{x}\rangle \mapsto \sum_{\boldsymbol{y}\in\mathbb{F}_2^m} (-1)^{\boldsymbol{x}\cdot\boldsymbol{y}} |\boldsymbol{y}\rangle .$$

The elements $\boldsymbol{x} = (x_1, x_2, \ldots, x_m) \in \mathbb{F}_2^m$ have a very natural quantum representation by m qubits:

$$|\boldsymbol{x}\rangle = |x_1\rangle |x_2\rangle \cdots |x_m\rangle ,$$

which corresponds exactly to the algebraic decomposition

$$\mathbb{F}_2^m = \mathbb{F}_2 \oplus \ldots \oplus \mathbb{F}_2.$$

Thus, it follows from Lemma 4.1.1 that it suffices to find QFT on $\mathbb{F}_2$, and the m-fold tensor product of that mapping will perform QFT on $\mathbb{F}_2^m$. However, QFT on $\mathbb{F}_2$ with representation $\{|0\rangle, |1\rangle\}$ is defined by

$$|0\rangle \mapsto \frac{1}{\sqrt{2}}(|0\rangle + |1\rangle),$$
$$|1\rangle \mapsto \frac{1}{\sqrt{2}}(|0\rangle - |1\rangle),$$

which can be implemented by the *Hadamard matrix* (which is also denoted by W_2)

$$H = \frac{1}{\sqrt{2}} \begin{pmatrix} 1 & 1 \\ 1 & -1 \end{pmatrix} .$$

Thus, we have obtained the following result:

Lemma 4.1.2. *Let $H_m = H\otimes\cdots\otimes H$ (m times). Then for $|\boldsymbol{x}\rangle = |x_1\rangle \cdots |x_m\rangle$*

$$H_m |\boldsymbol{x}\rangle = \frac{1}{\sqrt{2^m}} \sum_{\boldsymbol{y}\in\mathbb{F}_2^m} (-1)^{\boldsymbol{x}\cdot\boldsymbol{y}} |\boldsymbol{y}\rangle . \tag{4.10}$$

For any m, the matrix H_m is also called a *Hadamard matrix.* QFT (4.10) is also called a *Hadamard transform, Walsh transform*, or *Hadamard-Walsh transform.*

4.1.3 Quantum Fourier Transform in $\mathbb{Z}_n$

All the characters of group $\mathbb{Z}_n$ are

$$\chi_y(x) = e^{\frac{2\pi i x y}{n}},$$

where x and y are some representatives of cosets (See Sections 9.1.3, 9.1.4, and 9.2). To simplify the notations, we will denote any coset $k+n\mathbb{Z}$ by number the k. Using these notations,

$$\mathbb{Z}_n = \{0, 1, 2, \ldots, n-1\},$$

where addition is computed modulo n. This notation will be used hereafter.

The corresponding QFT on $\mathbb{Z}_n$ is the operation

$$|x\rangle \rightarrow \sum_{y=0}^{n-1} e^{-\frac{2\pi ixy}{n}} |y\rangle . \tag{4.11}$$

Now we have the problem that, unlike for $\mathbb{F}_2^m$, there is no evident way to give a quantum representation for the elements of $\mathbb{Z}_n$ in such a way that the basis

$$|0\rangle, |1\rangle, |2\rangle, \ldots, |n-1\rangle$$

could be represented as a tensor product of two smaller bases representing some subsystems of $\mathbb{Z}_n$. If, however, we know some factorization $n = n_1 n_2$ such that $\gcd(n_1, n_2) = 1$, the Chinese Remainder Theorem (Theorem 9.1.2) offers us a decomposition: according to the Chinese Remainder Theorem, there is a bijection

$$F : \mathbb{Z}_{n_1} \times \mathbb{Z}_{n_2} \rightarrow \mathbb{Z}_n$$

given by $F((k_1, k_2)) = a_1 n_2 k_1 + a_2 n_1 k_2$, where a_1 (resp. a_2) is the multiplicative inverse of n_2 (resp. n_1) modulo n_1 (resp. n_2). By Exercise 2, the mapping F is, in fact, an isomorphism. We also notice that, since a_1 (resp. a_2) has multiplicative inverse, the mapping $k_1 \mapsto a_1 k_1$ (resp. $k_2 \mapsto a_2 k_2$) is a permutation of $\mathbb{Z}_{n_1}$ (resp. $\mathbb{Z}_{n_2}$), and this is all we need for decomposing the QFT (4.11).

Assume that the QFTs are available for $\mathbb{Z}_{n_1}$ and $\mathbb{Z}_{n_2}$, i.e., we have quantum representations

$$\{|0\rangle, |1\rangle, \ldots, |n_1 - 1\rangle\} \text{ and } \{|0\rangle, |1\rangle, \ldots, |n_2 - 1\rangle\}$$

of $\mathbb{Z}_{n_1}$ and $\mathbb{Z}_{n_2}$ and we also have the routines for mappings

$$|k_1\rangle \mapsto \frac{1}{\sqrt{n_1}} \sum_{l_1=0}^{n_1-1} e^{-\frac{2\pi i l_1 k_1}{n_1}} |l_1\rangle \text{ and } |k_2\rangle \mapsto \frac{1}{\sqrt{n_2}} \sum_{l_2=0}^{n_2-1} e^{-\frac{2\pi i l_2 k_2}{n_2}} |l_2\rangle .$$

We use the quantum representation $|k\rangle = |k_1\rangle |k_2\rangle$ for the elements of $\mathbb{Z}_n$ given by the Chinese Remainder Theorem with $k = a_1 n_2 k_1 + a_2 n_1 k_2$. By constructing a quantum circuit that performs the multiplications $(k_1, k_2) \mapsto (a_1 k_1, a_2 k_2)$ and concatenating that with the QFT circuits on both components, we get

$$
\begin{aligned}
|k_1\rangle\,|k_2\rangle &\mapsto |a_1k_1\rangle\,|a_2k_2\rangle \\
&\mapsto \Big(\frac{1}{\sqrt{n_1}}\sum_{l_1=0}^{n_1-1} e^{-\frac{2\pi i l_1 a_1 k_1}{n_1}}\,|l_1\rangle\Big)\Big(\frac{1}{\sqrt{n_2}}\sum_{l_2=0}^{n_2-1} e^{-\frac{2\pi i l_2 a_2 k_2}{n_2}}\,|l_2\rangle\Big) \\
&= \frac{1}{\sqrt{n_1n_2}}\sum_{l_1=0}^{n_1-1}\sum_{l_2=0}^{n_2-1} e^{-\frac{2\pi i(a_1n_2l_1k_1+a_2n_1l_2k_2)}{n_1n_2}}\,|l_1\rangle\,|l_2\rangle \\
&= \frac{1}{\sqrt{n_1n_2}}\sum_{l_1=0}^{n_1-1}\sum_{l_2=0}^{n_2-1} e^{-\frac{2\pi i F(l_1k_1,l_2k_2)}{n_1n_2}}\,|l_1\rangle\,|l_2\rangle \\
&= \frac{1}{\sqrt{n_1n_2}}\sum_{l_1=0}^{n_1-1}\sum_{l_2=0}^{n_2-1} e^{-\frac{2\pi i F(k_1,k_2)F(l_1,l_2)}{n_1n_2}}\,|F(l_1,l_2)\rangle\,.
\end{aligned}
$$

This decomposition can be applied recursively to n_1 and n_2.

Notice carefully that the previous decomposition requires some factorization $n = n_1n_2$ such that $\gcd(n_1, n_2) = 1$, but generally the problem is to find a nontrivial factorization of n. Shor's factorization algorithm makes use of the inverse QFT on $\mathbb{Z}_{2^m}$, and since there is no coprime factorization for 2^m, we have to be ready to offer something else. Fortunately, we can also obtain some more knowledge on QTFs on groups $\mathbb{Z}_{2^m}$.

Since the inverse Fourier transform in $\mathbb{Z}_n$ is quite symmetric to (4.11), ($e^{-\frac{2\pi ixy}{n}}$ is replaced with $e^{\frac{2\pi ixy}{n}}$), choosing which one to study is a matter of personal preference. In what follows, we will learn how to implement the inverse Fourier transform on $\mathbb{Z}_{2^m}$. Group $\mathbb{Z}_{2^m}$ has a very natural representation by using m qubits: an element $x = x_{m-1}2^{m-1} + x_{m-2}2^{m-2} + \ldots + x_1 2 + x_0$, where each $x_i \in \{0, 1\}$, is represented as

$$
|x\rangle = |x_{m-1}\rangle\,|x_{m-2}\rangle \cdots |x_1\rangle\,|x_0\rangle\,.
$$

How do to implement the inverse QFT

$$
|x\rangle \mapsto \frac{1}{\sqrt{2^m}}\sum_{y=0}^{2^m-1} e^{\frac{2\pi ixy}{2^m}}\,|y\rangle \tag{4.12}
$$

with a quantum circuit? To approach this problem, we can first make an observation which should no longer be a surprise more after the previous decomposition results: The superposition on the right-hand side of (4.12) is decomposable.

Lemma 4.1.3.

$$
\begin{aligned}
&\sum_{y=0}^{2^m-1} e^{\frac{2\pi ixy}{2^m}}\,|y\rangle \\
&= (|0\rangle + e^{\frac{\pi ix}{2^0}}\,|1\rangle)(|0\rangle + e^{\frac{\pi ix}{2^1}}\,|1\rangle)\cdots(|0\rangle + e^{\frac{\pi ix}{2^{m-1}}}\,|1\rangle)
\end{aligned} \tag{4.13}
$$

Proof. Representing $|y\rangle = |y'b\rangle = |y'\rangle\,|b\rangle$, where y' are the $m-1$ most significant bits and b the least significant bit of y, we can divide the sum into two parts:

$$\begin{aligned}
\sum_{y=0}^{2^m-1} e^{\frac{2\pi i x y}{2^m}}\,|y\rangle &= \sum_{y'=0}^{2^{m-1}-1} e^{\frac{2\pi i x\cdot 2y'}{2^m}}\,|y'0\rangle + \sum_{y'=0}^{2^{m-1}-1} e^{\frac{2\pi i x(2y'+1)}{2^m}}\,|y'1\rangle \\
&= \sum_{y=0}^{2^{m-1}-1} e^{\frac{2\pi i 2xy}{2^m}}\,|y\rangle\,|0\rangle + \sum_{y=0}^{2^{m-1}-1} e^{\frac{2\pi i 2xy}{2^m}} e^{\frac{2\pi i x}{2^m}}\,|y\rangle\,|1\rangle \\
&= \sum_{y=0}^{2^{m-1}-1} e^{\frac{2\pi i x y}{2^{m-1}}}\,|y\rangle\,(|0\rangle + e^{\frac{\pi i x}{2^{m-1}}}\,|1\rangle).
\end{aligned}$$

The claim now follows by induction. □

Decomposition (4.13) also gives us a clue on how to compute transform (4.12) by using a quantum circuit. The lth phase of $|1\rangle$ in (4.13) depends only on bits $x_0, x_1, \ldots, x_{l-1}$ and can be written as

$$\begin{aligned}
&\exp\left(\frac{\pi i(2^{m-1}x_{m-1} + 2^{m-2}x_{m-2} + \ldots + 2x_1 + x_0)}{2^{l-1}}\right) \\
&= \exp\left(\frac{\pi i(2^{l-1}x_{l-1} + 2^{l-2}x_{l-2} + \ldots + 2x_1 + x_0)}{2^{l-1}}\right) \\
&= \exp\left(\frac{\pi i x_{l-1}}{2^0}\right)\exp\left(\frac{\pi i x_{l-2}}{2^1}\right)\cdots\exp\left(\frac{\pi i x_1}{2^{l-2}}\right)\exp\left(\frac{\pi i x_0}{2^{l-1}}\right) \\
&= (-1)^{x_{l-1}}\exp\left(\frac{\pi i x_{l-2}}{2^1}\right)\cdots\exp\left(\frac{\pi i x_1}{2^{l-2}}\right)\exp\left(\frac{\pi i x_0}{2^{l-1}}\right)
\end{aligned}$$

We can now describe a quantum circuit that performs operation (4.12): given a quantum representation

$$|x\rangle = |x_{m-1}\rangle\,|x_{m-2}\rangle\cdots|x_1\rangle\,|x_0\rangle \tag{4.14}$$

we swap the qubits to get the *reverse binary representation*

$$|x\rangle = |x_0\rangle\,|x_1\rangle\cdots|x_{m-2}\rangle\,|x_{m-1}\rangle\,. \tag{4.15}$$

In practice, this swapping is not necessary, since we could operate as well with the inverse binary representations; we included the swapping here just for the sake of completeness.

On state (4.15) we proceed from right to left as follows: Hadamard transform on the mth (the right-most) qubit gives

$$\frac{1}{\sqrt{2}}\,|x_0\rangle\,|x_1\rangle\cdots|x_{m-2}\rangle\,(|0\rangle + (-1)^{x_{m-1}}\,|1\rangle),$$

and we complete the phase $(-1)^{x_{m-1}}$ to

$$(-1)^{x_{m-1}} \exp(\frac{\pi i x_{m-2}}{2^1}) \cdots \exp(\frac{\pi i x_1}{2^{m-2}}) \exp(\frac{\pi i x_0}{2^{m-1}})$$

by using phase rotations

$$\exp(\frac{\pi i}{2^1}), \ldots, \exp(\frac{\pi i}{2^{m-2}}), \exp(\frac{\pi i}{2^{m-1}})$$

conditionally. That is, for each $l \in \{1, 2, \ldots, m-1\}$, a phase factor $\exp(\frac{\pi i}{2^{m-l}})$ is introduced to the mth bit if and only if mth and lth qubits are both 1. This procedure will yield state

$$\frac{1}{\sqrt{2}} |x_0\rangle |x_1\rangle \cdots |x_{m-2}\rangle (|0\rangle + e^{\frac{2\pi i x}{2^{m-1}}} |1\rangle).$$

The same procedure will be applied from right to left to qubits at locations $m-1$, ..., 2, and 1: for a qubit at location l we first perform a Hadamard transform to get the qubit in state

$$\frac{1}{\sqrt{2}}(|0\rangle + (-1)^{x_{l-1}} |1\rangle),$$

then, for each k in range $l-1$, $l-2$, ..., 1, we introduce phase factor

$$\exp(\frac{\pi i}{2^{l-k}})$$

conditionally if and only if the lth and kth qubit are both 1. That can be achieved by mapping

$$\begin{aligned}
&|0\rangle |0\rangle \mapsto |0\rangle |0\rangle \\
&|0\rangle |1\rangle \mapsto |0\rangle |1\rangle \\
&|1\rangle |0\rangle \mapsto |1\rangle |0\rangle \\
&|1\rangle |1\rangle \mapsto e^{\frac{\pi i}{2^{l-k}}} |1\rangle |1\rangle ,
\end{aligned}$$

which acts on the lth and kth qubits. The matrix of the mapping can be written as

$$\phi_{k,l} = \begin{pmatrix} 1 & 0 & 0 & 0 \\ 0 & 1 & 0 & 0 \\ 0 & 0 & 1 & 0 \\ 0 & 0 & 0 & e^{\frac{\pi i}{2^{l-k}}} \end{pmatrix}. \tag{4.16}$$

Ignoring the swapping $(x_{m-1}, x_{m-2}, \ldots, x_0) \mapsto (x_0, x_1, \ldots, x_{m-1})$, this procedure results in the network in Figure 4.1 with $\frac{1}{2}m(m+1)$ gates.

In Figure 4.1, the subindices of the gates ϕ are omitted for typographical reasons. The ϕ-gates are (from left to right) $\phi_{m-1,m-2}$, $\phi_{m-1,m-3}$, $\phi_{m-1,0}$, $\phi_{m-2,m-3}$, $\phi_{m-2,0}$, and $\phi_{m-3,0}$.

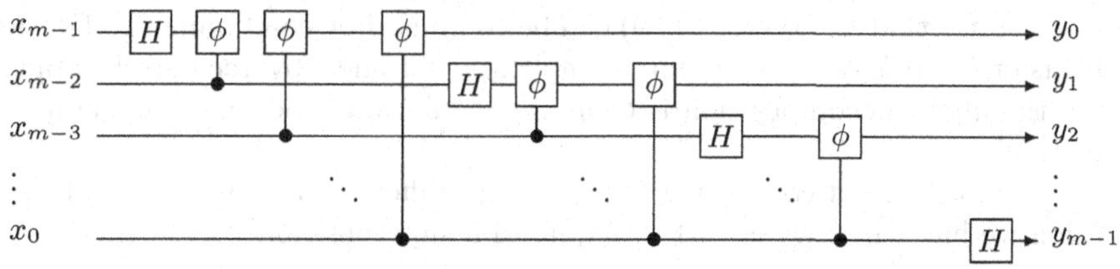

Fig. 4.1. Quantum network for QFT on $\mathbb{Z}_{2^m}$

4.1.4 Complexity Remarks

By the traditional discrete Fourier transform one usually understands the Fourier transform in $\mathbb{Z}_n$ given by

$$\widehat{f}(x) = \sum_{y=0}^{n-1} e^{-\frac{2\pi i x y}{n}} f(y). \tag{4.17}$$

For practical reasons, n is typically chosen as $n = 2^m$. Fourier transform (4.17) can be used to approximate the continuous Fourier transform, and has therefore tremendous importance in physics and engineering.

By computing (4.17), we understand that we are given vector

$$(f(0), f(1), \ldots, f(2^m - 1)) \tag{4.18}$$

as the input data, and the required output is

$$(\widehat{f}(0), \widehat{f}(1), \ldots, \widehat{f}(2^m - 1)). \tag{4.19}$$

The naive method for computing the Fourier transform is to use formula (4.17) straightforwardly to find all elements in (4.19), and the time complexity given by this method is $O((2^m)^2)$, as one can easily see.

A significant improvement is obtained by using fast Fourier transform (FFT), whose core can be expressed in a decomposition

$$\begin{aligned}
\widehat{f}(x) &= \frac{1}{\sqrt{2^m}} \sum_{y=0}^{2^m-1} e^{-\frac{2\pi i x y}{2^m}} f(y) \\
&= \frac{1}{\sqrt{2^m}} \Big(\sum_{y=0}^{2^{m-1}-1} e^{-\frac{2\pi i x \cdot 2y}{2^m}} f(2y) + \sum_{y=0}^{2^{m-1}-1} e^{-\frac{2\pi i x \cdot (2y+1)}{2^m}} f(2y+1) \Big) \\
&= \frac{1}{\sqrt{2^m}} \Big(\sum_{y=0}^{2^{m-1}-1} e^{-\frac{2\pi i x y}{2^{m-1}}} f(2y) + e^{-\frac{2\pi i x}{2^m}} \sum_{y=0}^{2^{m-1}-1} e^{-\frac{2\pi i x y}{2^{m-1}}} f(2y+1) \Big),
\end{aligned}$$

which very closely resembles the decomposition of Lemma 4.1.3, and essentially states that the vector (4.19) can be computed by combining two Fourier transforms in $\mathbb{Z}_{2^{m-1}}$. The time complexity obtained by recursively applying the above decomposition is $O(m2^m)$, significantly better than the naive method.

The problem of computing QFT is quite different: in a very typical situation we have, instead of (4.18), a quantum superposition

$$c_0 |0\rangle + c_1 |1\rangle + \ldots + c_{2^m-1} |2^m - 1\rangle \tag{4.20}$$

and QFT operates on the *coefficients* of (4.20). At the same time, the physical representation size of (4.20) is small; in system (4.20) there are only m qubits, yet there are 2^m coefficients c_i. Earlier we learned that QFT in $\mathbb{Z}_{2^m}$ can be done in time $O(m^2)$ (Hadamard-Walsh transform can be done even in time $O(m)$), which is exponentially separate from the classical counterparts of Fourier transform. But the major difference is that the physical representation sizes of (4.18) and (4.20) are also exponentially separate, the first one taking $\Omega(2^m)$ bits and the latter one m qubits. Later, we will learn that the key idea behind many interesting fast quantum algorithms is the use of quantum parallelism to convey some information of interest into the coefficients of (4.20) and then compute QFT rapidly.

4.2 Shor's Algorithm for Factoring Numbers

4.2.1 From Periods to Factoring

Given two prime numbers p and q, it is an easy task to compute the product $n = pq$. The naive algorithm already has quadratic time complexity $O(\max\{|p|, |q|\}^2)$, but, by using more sophisticated methods, even performance $O(\max\{|p|, |q|\}^{1+\epsilon})$ is reachable for any $\epsilon > 0$ (see [28]). On the other hand, the inverse problem, factorization, seems to be extremely hard to solve.

Remark 4.2.1. A deterministic polynomial-time algorithm for recognizing prime numbers has been recently discovered [1].

Nowadays it is a widespread belief that there is no efficient classical algorithm for solving the factorization problem: given $n = pq$, a product of two primes, find out p and q. This problem has great importance in cryptography: the reliability of the famous public-key cryptosystem RSA is based on the assumption that no efficient algorithm exists for factorization.

In 1994, Peter W. Shor [81] introduced a probabilistic polynomial-time quantum algorithm for factorization, which will be presented in this chapter. First we show how to reduce the factoring to finding orders of elements in $\mathbb{Z}_n$. Let

$$n = p_1^{e_1} \cdots p_k^{e_k} \tag{4.21}$$

be the (unknown) prime factorization of an odd number (the powers of 2 can be easily recognized and canceled). We will also assume that $k \geq 2$, i.e., n has at least two different prime factors. In fact, there is a polynomial-time algorithm for recognizing powers (see Exercise 3), so we may as well assume that n is not a prime power. We can be satisfied if we can find any non-trivial factor of n rapidly, since the process can be recursively applied to all of the factors in order to uncover the whole factorization (4.21) swiftly.

Let us randomly choose, with uniform probability, an element $a \in \mathbb{Z}_n$, $a \neq 1$. If $d = \gcd(a, n) > 1$, then d is a nontrivial factor of n and our task is complete. If $d = 1$, assume that we could somehow extract $r = \text{ord}_n(a)$ ($\text{ord}_n(a)$ means the order of element a in $\mathbb{Z}_n$, see Sections 9.1.3 and 9.1.4 for the number-theoretic notions in this section). Then

$$a^r \equiv 1 \pmod{n},$$

which means that n divides $a^r - 1$. Of course this does not yet offer us a method for extracting a nontrivial factor of n, but if r is even, we can easily factorize $a^r - 1$:

$$a^r - 1 = (a^{\frac{r}{2}} - 1)(a^{\frac{r}{2}} + 1).$$

Now, since n divides $a^r - 1$, n must share a factor with $a^{\frac{r}{2}} - 1$ or with $a^{\frac{r}{2}} + 1$ (or with both), and this factor can be extracted by using Euclid's algorithm.

How could we guarantee that the factor of n obtained in this way is nontrivial? Namely, if n divides $a^{\frac{r}{2}} \pm 1$, then Euclid's algorithm would only give n as the greatest common divisor. Fortunately, it is not so likely that n divides $a^{\frac{r}{2}} \pm 1$, as we will soon demostrate. First, $n \mid (a^{\frac{r}{2}} - 1)$ would imply that

$$a^{\frac{r}{2}} \equiv 1 \pmod{n},$$

which would be absurd by the very definition of order. But it may still happen that n just divides $a^{\frac{r}{2}} + 1$, and does not share any factor with $a^{\frac{r}{2}} - 1$. In this case, the factor given by Euclid's algorithm would also be n. On the other hand, $n \mid a^{\frac{r}{2}} + 1$ means that

$$a^{\frac{r}{2}} \equiv -1 \pmod{n},$$

i.e., $a^{\frac{r}{2}}$ is a nontrivial square root of 1 modulo n. In the next section we will use elementary number theory to show that, for a randomly chosen (with uniform distribution) element $a \in \mathbb{Z}_n^*$, the probability that $r = \text{ord}(a)$ is even, and that $a^{\frac{r}{2}} \not\equiv -1 \pmod{n}$ is at least $\frac{1}{2}$. Consequently, assuming that $r = \text{ord}(a)$ could be rapidly extracted, we could, with a reasonable probability, find a nontrivial factor of n.

There may be still some concern about the magnitude of numbers $a^{\frac{r}{2}} \pm 1$. Is it possible that these numbers would be so large that the procedure could be inefficient anyway? Fortunately, it is easy to answer this question: divisibility by n is a periodic property with period n; so, once $r = \text{ord}(a)$ is found, it suffices to find $a^{\frac{r}{2}} \pm 1$ *modulo* n (for modular exponentiation, rapid algorithms are known).

We may now focus on the problem of finding $r = \text{ord}(a)$. Since n is odd but not a prime power, group $\mathbb{Z}_n^*$ is never cyclic, but each element $a \in \mathbb{Z}_n^*$ has an order $r = \text{ord}(a)$ smaller than $\frac{1}{2}\varphi(n) < \frac{n}{2}$. By the definition of the order,

$$a^{l+sr} \equiv a^l \pmod{n}$$

for any integer s, which implies that the function $f : \mathbb{Z} \to \mathbb{Z}_n$ defined by

$$f(k) = a^k \pmod{n} \tag{4.22}$$

has r as the period. We will demonstrate in Section 4.2.3 that it is possible to find this period rapidly by using QFT.

Example 4.2.1. $15 = 3 \cdot 5$ is the smallest number that can be factorized by the above method.

$$\mathbb{Z}_{15}^* = \{1, 2, 4, 7, 8, 11, 13, 14\},$$

and the orders of the elements are 0, 4, 2, 4, 4, 2, 4, 2 respectively. The only element for which $a^{\frac{r}{2}} \equiv -1 \pmod{15}$ is $a = 14 \equiv -1 \pmod{15}$. If, for example, $a = 7$, then $15 \mid 7^4 - 1$ and $7^2 - 1 \equiv 3 \pmod{15}$, $7^2 + 1 \equiv 5 \pmod{15}$. We get $3 = \gcd(3, 15)$ and $5 = \gcd(5, 15)$ as nontrivial factors of 15.

4.2.2 Orders of the Elements in $\mathbb{Z}_n$

Let $n = p_1^{e_1} \cdots p_k^{e_k}$ be the prime factorization of an odd n, and $a \in \mathbb{Z}_n^*$ a randomly (and uniformly) chosen element. The Chinese Remainder Theorem gives us a useful decomposition

$$\mathbb{Z}_n^* = \mathbb{Z}_{p_1^{e_1}}^* \times \cdots \times \mathbb{Z}_{p_k^{e_k}}^* \tag{4.23}$$

of $\mathbb{Z}_n^*$ as a direct product of cyclic groups. Recall that the cardinality of each factor is given by $\left|\mathbb{Z}_{p_i^{e_i}}\right| = \varphi(p_i^{e_i}) = p_i^{e_i - 1}(p_i - 1)$ (see Section 9.1.4). Note especially that each $\mathbb{Z}_{p_i^{e_i}}$ has an even cardinality. For the decomposition of an element $a \in \mathbb{Z}_n^*$ we will use notation

$$(a_1, \ldots, a_k) \in \mathbb{Z}_{p_1^{e_1}}^* \times \ldots \times \mathbb{Z}_{p_k^{e_k}}^*, \tag{4.24}$$

and the order of an element a in $\mathbb{Z}_{p^e}^*$ will be denoted by $\text{ord}_{p^e}(a)$.

Because of the decomposition (4.24), to choose a random element $a \in \mathbb{Z}_n^*$ is to choose k random elements as in (4.24) and vice versa. The following technical lemma will be quite useful for the probability estimations.

Lemma 4.2.1. *Let $\varphi(p^e) = 2^u v$, where $u \geq 1$, $2 \nmid v$ and $s \geq 0$ a fixed integer. Then the probability that a randomly (and uniformly) chosen element $a \in \mathbb{Z}_{p^e}^*$ has an order of form $2^s t$ with $2 \nmid t$ is at most $\frac{1}{2}$.*

Proof. If $s > u$, then the probability in question is 0, so we may assume $s \leq u$. Let g be a generator of $\mathbb{Z}_{p^e}^*$. Then $\mathbb{Z}_{p^e}^* = \{g^0, g^1, \ldots, g^{2^u v - 1}\}$, and

$$\mathrm{ord}(g^j) = \frac{2^u v}{\gcd(j, 2^u v)},$$

so an order of form $2^s t$ occurs if and only if $j = 2^{u-s} w$, where $2 \nmid w$. In the set $\{0, 1, 2, \ldots, 2^u v - 1\}$ there are exactly $2^s v$ multiples of 2^{u-s}, namely $0 \cdot 2^{u-s}$, $1 \cdot 2^{u-s}$, ..., and $(2^s v - 1)2^{u-s}$, but only half of them have an odd multiplier. Therefore, the required probability is

$$\frac{\frac{1}{2} \cdot 2^s v}{2^u v} = \frac{1}{2} \frac{2^s}{2^u} \leq \frac{1}{2},$$

as claimed. □

We can use the previous lemma to estimate the probability of having an odd order.

Lemma 4.2.2. *The probability that $r = \mathrm{ord}(a)$ is odd for a uniformly chosen $a \in \mathbb{Z}_n^*$ is at most $\frac{1}{2^k}$.*

Proof. Let $(a_1, \ldots, a_k)$ be the decomposition (4.24) of an element $a \in \mathbb{Z}_n^*$. Let $r_i = \mathrm{ord}_{p_i^{e_i}}(a_i)$. Since $r = \mathrm{lcm}\{r_1, \ldots, r_k\}$ (Exercise 4), r is odd if and only if each r_i is odd.

Putting $s = 0$ in Lemma 4.2.1, we learn that the probability of having a random $a_i \in \mathbb{Z}_{p_i^{e_i}}$ with odd order is at most $\frac{1}{2}$. Therefore, the probability P_1 of having odd r is at most

$$P_1 \leq \frac{1}{2} \cdots \frac{1}{2} = \frac{1}{2^k},$$

as claimed. □

What about the probability that $a^{\frac{r}{2}} \equiv -1 \pmod{n}$? It turns out that Lemma 4.2.1 is useful for estimating that probability, too.

Lemma 4.2.3. *Let $n = p_1^{e_1} \cdots p_k^{e_k}$ be the prime decomposition of an odd n and $k \geq 2$. If $r = \mathrm{ord}_n(a)$ is even, then the probability that $a^{\frac{r}{2}} \equiv -1 \pmod{n}$ is at most $\frac{1}{2^{k-1}}$.*

Proof. Congruence $a^{\frac{r}{2}} \equiv -1 \pmod{n}$ implies that

$$a^{\frac{r}{2}} \equiv -1 \pmod{p_i^{e_i}} \tag{4.25}$$

for each i. Let $r_i = \mathrm{ord}_{p_i^{e_i}}(a_i)$, so $r = \mathrm{lcm}\{r_1, \ldots, r_k\}$. We write $r = 2^s t$ and $r_i = 2^{s_i} t_i$, where $2 \nmid t$ and $2 \nmid t_i$. From the fact that $r_i \mid r$ for each i, it follows that $s_i \leq s$, but the congruencies (4.25) can only hold if $s_i = s$ for every i. For if $s_i < s$ for some i, then also $r_i \mid \frac{r}{2}$ (the assumption $k \geq 2$ is needed here!), which implies that

$$a^{\frac{r}{2}} \equiv 1 \pmod{p_i^{e_i}}. \tag{4.26}$$

But (4.26) together with (4.25) gives $1 \equiv -1 \pmod{p_i^{e_i}}$, which is absurd since $p_i \neq 2$.

Therefore, the probability P_2 that $a^{\frac{r}{2}} \equiv -1 \pmod{n}$ is at most the probability that $s_i = s$ for each i, which is

$$\begin{aligned} P_2 &= \sum_{l=0}^{\infty} P(s_1 = l) \prod_{i=2}^{k} P(s_i = s_1) \\ &= \prod_{i=2}^{k} P(s_i = s_1) \sum_{l=0}^{\infty} P(s_1 = l) \\ &= \prod_{i=2}^{k} P(s_i = s) \leq \frac{1}{2^{k-1}} \end{aligned}$$

by Lemma 4.2.1. □

Combining Lemmata 4.2.2 and 4.2.3, we get the following lemma.

Lemma 4.2.4. *Let $n = p_1^{e_1} \cdots p_k^{e_k}$ be the prime factorization of an odd n with $k \geq 2$. Then, for a random $a \in \mathbb{Z}_n^*$ (chosen uniformly), the probability that $r = \mathrm{ord}_n(a)$ is even and $a^{\frac{r}{2}} \not\equiv -1 \pmod{n}$ is at least*

$$(1 - \frac{1}{2^k})(1 - \frac{1}{2^{k-1}}) \geq \frac{3}{4} \cdot \frac{1}{2} = \frac{3}{8}.$$

For our purposes, the result of the previous lemma would be sufficient. On the other hand, it would be theoretically interesting to see if the result could be improved. In fact, this is the case:

Lemma 4.2.5. *Let all the notations be as before. Then the probability that $r = \mathrm{ord}_n(a)$ is odd or $a^{\frac{r}{2}} \equiv -1 \pmod{n}$ is at most $\frac{1}{2^{k-1}}$.*

Proof. Recalling the previous notations, the result follows directly from the observation that r is odd if and only if each $s_i = 0$, and that if $a^{\frac{r}{2}} \equiv -1 \pmod{n}$ happens, then all numbers s_i are equal. The former event is a subcase of the latter, so we may conclude that

$$\begin{aligned} &P(r \text{ is odd or } a^{\frac{r}{2}} \equiv -1 \pmod{n}) \\ &\leq P(\text{all numbers } s_i \text{ are equal}) \leq \frac{1}{2^{k-1}}. \end{aligned}$$

The latter inequality follows directly from that one of Lemma 4.2.3. □

Corollary 4.2.1. *Let all the notations be as in Lemma 4.2.4. The probability that that r is even and $a^{\frac{r}{2}} \not\equiv -1 \pmod{n}$ is at least $\frac{1}{2}$.*

Remark 4.2.2. By studying the group $\mathbb{Z}_{21}^*$ we can see that the probability limit of the previous lemma is optimal.

4.2.3 Finding the Period

By Lemma 4.2.4 we know that, for a randomly (and uniformly) chosen $a \in \mathbb{Z}_n^*$, the probability that $r = \text{ord}_n(a)$ is even and $a^{\frac{r}{2}} \not\equiv -1$ is at least $\frac{9}{16}$ for any odd n having at least two distinct prime factors. It was already discussed in Section 4.2.1 that the knowledge about the order of such an element a allows us to find efficiently a nontrivial factor of n. Thus, a procedure for computing the order would provide an efficient probabilistic method for finding the factors. Therefore, we have a good reason to believe that finding $r = \text{ord}_n(a)$ is computationally a very difficult task.

It is clear that finding $r = \text{ord}_n(a)$ reduces to finding the period of function $f : \mathbb{Z} \to \mathbb{Z}_n$ defined by

$$f(k) = a^k \pmod{n}. \tag{4.27}$$

In fact, if the period of f is r, then $a^r = a^0 \equiv 1 \pmod{n}$.

It is a well-known fact that Fourier transform can be used for extracting information about the periodicity (see Section 9.2.5), and we will show that, by using QFT, the period of (4.27) can be found with non-negligible probability.

Of course, we cannot compute the QFT of (4.27) on whole $\mathbb{Z}$, so we choose a domain $\mathbb{Z}_m = \{0, 1, \ldots, m-1\}$ with $m > n$ large enough to give room for the period to appear. We will also choose $m = 2^l$ for two reasons: the first is that then there is a natural quantum representation of elements of $\mathbb{Z}_m$ by l qubits; and the second is that we have already learned in Section 4.1.3 how to compute the QFT in $\mathbb{Z}_{2^l}$. We will fix this l later.

The procedure requires quantum representation of $\mathbb{Z}_m$ and $\mathbb{Z}_n$ (the latter can be replaced by some set including $\mathbb{Z}_n$). The first takes l qubits and the other takes at most l qubits. For the representation, notation $|x\rangle\,|y\rangle$, where $x \in \mathbb{Z}_m$, $y \in \mathbb{Z}_n$, will be used.

Quantum Algorithm for Finding the Orders

1. The first stage is to prepare a superposition

$$\frac{1}{\sqrt{m}} \sum_{k=0}^{m-1} |k\rangle\,|0\rangle \tag{4.28}$$

by beginning with $|0\rangle\,|0\rangle$ and using Hadamard-Walsh transform in $\mathbb{Z}_m$.

2. The second step is to compute $k \mapsto a^k \pmod{n}$ to get

$$\frac{1}{\sqrt{m}} \sum_{k=0}^{m-1} |k\rangle \left|a^k\right\rangle . \tag{4.29}$$

Since the function $k \mapsto a^k \pmod{n}$ has period r, (4.29) can be written as

$$\frac{1}{\sqrt{m}} \sum_{l=0}^{r-1} \sum_{q=0}^{s_l} |qr+l\rangle \left|a^l\right\rangle , \tag{4.30}$$

where s_l is the greatest integer for which $s_l r + l < m$. It is clear that s_l cannot vary very much: we always have $\frac{m}{r} - 1 - \frac{l}{r} \le s_l < \frac{m}{r} - \frac{l}{r}$.
3. Compute the inverse QFT on $\mathbb{Z}_m$ to get

$$\begin{aligned} &\frac{1}{\sqrt{m}} \sum_{l=0}^{r-1} \sum_{q=0}^{s_l} \frac{1}{\sqrt{m}} \sum_{p=0}^{m-1} e^{\frac{2\pi i p(qr+l)}{m}} |p\rangle \left|a^l\right\rangle \\ &= \frac{1}{m} \sum_{l=0}^{r-1} \sum_{p=0}^{m-1} e^{\frac{2\pi i p l}{m}} \sum_{q=0}^{s} e^{\frac{2\pi i p r q}{m}} |p\rangle \left|a^l\right\rangle . \end{aligned} \tag{4.31}$$

4. Observe the quantum representation of $\mathbb{Z}_m$ in superposition (4.31) to get some $p \in \mathbb{Z}_m$.
5. Find the convergents $\frac{p_i}{q_i}$ of the continuous fraction expansion (see Section 9.4.2) of $\frac{p}{m}$ using Euclid's algorithm and output the smallest q_i such that $a^{q_i} \equiv 1 \pmod{n}$, if such q_i exists.

In the next section we will estimate the correctness probability of this method.

Example 4.2.2. Let $n = 15$ and $a = 7$ be the element whose period is to be found. Let us choose $m = 16$.

1. The first step is to prepare

$$\frac{1}{4} \sum_{k=0}^{15} |k\rangle |0\rangle .$$

2. Computation of $k \mapsto 7^k \pmod{15}$ gives

$$\begin{aligned} &\frac{1}{4}\big(|0\rangle |1\rangle + |1\rangle |7\rangle + |2\rangle |4\rangle + |3\rangle |13\rangle + |4\rangle |1\rangle + \ldots + |15\rangle |13\rangle \big) \\ &= \frac{1}{4}\Big(\big(|0\rangle + |4\rangle + |8\rangle + |12\rangle \big) |1\rangle \\ &\quad + \big(|1\rangle + |5\rangle + |9\rangle + |13\rangle \big) |7\rangle \\ &\quad + \big(|2\rangle + |6\rangle + |10\rangle + |14\rangle \big) |4\rangle \\ &\quad + \big(|3\rangle + |7\rangle + |11\rangle + |15\rangle \big) |13\rangle \Big). \end{aligned}$$

3. The inverse QFT in $\mathbb{Z}_{16}$ yields

$$\frac{1}{4}\Big(\big(|0\rangle + |4\rangle + |8\rangle + |12\rangle\big)|1\rangle$$
$$+\big(|0\rangle + i\,|4\rangle - |8\rangle - i\,|12\rangle\big)|7\rangle$$
$$+\big(|0\rangle - |4\rangle + |8\rangle - |12\rangle\big)|4\rangle$$
$$+\big(|0\rangle - i\,|4\rangle - |8\rangle + i\,|12\rangle\big)|13\rangle\Big).$$

4. The probabilities to observe elements 0, 4, 8, and 12 are $\frac{1}{4}$ for each.
5. The only convergent of $\frac{0}{16}$ is just $\frac{0}{1}$, which does not give the period. The convergents of $\frac{4}{16}$ are $\frac{0}{1}$ and $\frac{1}{4}$, and the latter one gives the correct period 4. Observation of 8 does not give the period either, since the convergents of $\frac{8}{16}$ are $\frac{0}{1}$ and $\frac{1}{2}$, but 12 gives, the convergents of $\frac{12}{16}$ are $\frac{0}{1}$, $\frac{1}{1}$, and $\frac{3}{4}$.

4.3 The Correctness Probability

4.3.1 The Easy Case

The probability of seeing a particular $p \in \mathbb{Z}_m$ when observing (4.31) is

$$P(p) = \sum_{l=0}^{r-1} \frac{1}{m^2} \left| e^{\frac{2\pi i p l}{m}} \sum_{q=0}^{s_l} e^{\frac{2\pi i p q r}{m}} \right|^2 = \frac{1}{m^2} \sum_{l=0}^{r-1} \left| \sum_{q=0}^{s_l} e^{\frac{2\pi i p q r}{m}} \right|^2 . \tag{4.32}$$

We will show that, with a non-negligible probability, the value p received observing (4.31) will give us the period r.

To get some insight into how to evaluate the probability (4.32), we will first assume that r happens to divide m. Recalling that s_l is the greatest integer such that $s_l r + l < m$, and that $0 \le l < r$, we see that $s_l = \frac{m}{r} - 1$ for each l. In this case, (4.32) becomes

$$P(p) = \frac{r}{m^2} \left| \sum_{q=0}^{m/r-1} e^{\frac{2\pi i p q}{m/r}} \right|^2 . \tag{4.33}$$

The sum in (4.33) runs over all of the characters of $\mathbb{Z}_{m/r}$ evaluated at p. Thus (see Section 9.2.1),

$$\sum_{q=0}^{m/r-1} e^{\frac{2\pi i p q}{m/r}} = \begin{cases} \frac{m}{r}, & \text{if } p = 0 \text{ in } \mathbb{Z}_{m/r} \\ 0 & \text{otherwise,} \end{cases}$$

and therefore

$$P(p) = \begin{cases} \frac{1}{r}, & \text{if } p = d \cdot \frac{m}{r} \text{ for some } d \\ 0, & \text{otherwise.} \end{cases}$$

Thus, in the case $r \mid m$, observation of (4.31) can give only some p in the set $\{0 \cdot \frac{m}{r}, 1 \cdot \frac{m}{r}, \ldots, (r-1) \cdot \frac{m}{r}\}$, any such with a probability of $\frac{1}{r}$. Now that we know m and have learned $p = d \cdot \frac{m}{r}$ by observing (4.31), we may try to find r by canceling $\frac{p}{m} = \frac{d}{r}$ into an irreducible fraction using Euclid's algorithm. Unfortunately, this works for certain only if $\gcd(d, r) = 1$; in case $\gcd(d, r) > 1$ we will just get a factor of r as the nominator of $\frac{p}{m} = \frac{d}{r}$. Fortunately, we can show, by using the number-theoretical results, that the probability of having $\gcd(d, r) = 1$ does not converge to zero too fast.

Example 4.3.1. In Example 4.2.2, the period $r = 4$ divided $m = 16$, and the only elements that could be observed were 0, 4, 8, 12, the multiples of $4 = \frac{16}{4}$. However, 0 and 8 did not give the period, since the multipliers 0 and 2 share a factor with 4.

4.3.2 The General Case

Remembering that $m = 2^l$, it is clear that we cannot always be lucky enough to have a period r dividing m. We can, however, ask what is the probability of observing a p which is *close* to a multiple of $\frac{m}{r}$. For if

$$m\left|\frac{p}{m} - \frac{d}{r}\right| = \left|p - d\frac{m}{r}\right|$$

is small enough and $\gcd(d, r) = 1$, then $\frac{d}{r}$ is a convergent of $\frac{p}{m}$, and all the convergents of $\frac{p}{m}$ can be found efficiently by using Euclid's algorithm. (see Section 9.4.2).

We will now fix m in such a way that the continued fraction method will apply. For any integer $d \in \{0, 1, \ldots, r-1\}$, there is always a unique integer p such that the inequality

$$-\frac{1}{2} < p - d\frac{m}{r} \leq \frac{1}{2} \tag{4.34}$$

holds. Now choosing m such that $n^2 \leq m < 2n^2$ guarantees that

$$\left|\frac{p}{m} - \frac{d}{r}\right| \leq \frac{1}{2m} \leq \frac{1}{2n^2} < \frac{1}{2r^2},$$

which implies by Theorem 9.4.3 that if $\gcd(d, r) = 1$, then $\frac{d}{r}$ is a convergent to $\frac{p}{m}$.

But there are only r integers p in $\mathbb{Z}_m$ that satisfy (4.34), one for each $d \in \{0, 1, \ldots, r-1\}$. What is the probability of observing one? In other words, we should estimate the probability of observing p which satisfies

$$|pr - dm| \leq \frac{r}{2} \tag{4.35}$$

for some d.

Lemma 4.3.1. *For $n \geq 100$, the observation of (4.31) will give a $p \in \mathbb{Z}_m$ such that $|pr - dm| \leq \frac{r}{2}$ with a probability of not less than $\frac{2}{5}$.*

Proof. Evaluating (4.32) (Exercise 5), we get

$$P(p) = \frac{1}{m^2}\sum_{l=0}^{r-1}\frac{\sin^2\frac{\pi pr(s_l+1)}{m}}{\sin^2\frac{\pi pr}{m}}$$
$$= \frac{1}{m^2}\sum_{l=0}^{r-1}\frac{\sin^2\frac{\pi(pr-dm)(s_l+1)}{m}}{\sin^2\frac{\pi(pr-dm)}{m}}$$

for any fixed d by the periodicity of $\sin^2$. Since we now assume that $x = pr - dm$ takes values in $[-\frac{r}{2}, \frac{r}{2}]$, we will estimate

$$f(x) = \frac{\sin^2\frac{\pi x(s_l+1)}{m}}{\sin^2\frac{\pi x}{m}}.$$

It is not difficult to show that $f(x)$ is an even function, taking the maximum $(s_l+1)^2$ at $x = 0$ and minima in $[-\frac{r}{2}, \frac{r}{2}]$ at the end points $\pm\frac{r}{2}$. Therefore,

$$f(x) \geq \frac{\sin^2\frac{\pi r}{2m}(s_l+1)}{\sin^2\frac{\pi r}{2m}}.$$

Since s_l is the greatest integer such that $s_l r + l < m$,

$$\frac{\pi}{2}(1-\frac{r}{m}) < \frac{\pi r}{2m}(s_l+1) < \frac{\pi}{2}(1+\frac{r}{m}),$$

and

$$f(x) \geq \frac{\sin^2\frac{\pi}{2}(1-\frac{r}{m})}{\sin^2\frac{\pi r}{2m}}. \tag{4.36}$$

By the choice of $m \geq n^2$, $\frac{r}{m}$ is negligible, and we can use estimations $\sin x \leq x$ and $\sin^2\frac{\pi}{2}(1+x) \geq 1 - (\frac{\pi}{2}x)^2$ (Exercise 6) to get

$$f(x) \geq \frac{4}{\pi^2}\left(\frac{m}{r}\right)^2\left(1-\left(\frac{\pi}{2}\frac{r}{m}\right)^2\right), \tag{4.37}$$

which implies that

$$P(p) \geq \frac{4}{\pi^2}\frac{1}{r}\left(1-\left(\frac{\pi}{2}\frac{r}{m}\right)^2\right). \tag{4.38}$$

Factor

$$1-\left(\frac{\pi}{2}\frac{r}{m}\right)^2 \tag{4.39}$$

in (4.38) tends to 1 as $\frac{r}{m} \to 0$, and for $n \geq 100$, (4.39) is already greater than 0.9999. Thus, the probability of observing a fixed p such that $-\frac{1}{2} < p - d\frac{m}{r} \leq \frac{1}{2}$ is at least $\frac{2}{5}\frac{1}{r}$ for any $n \geq 100$.

But there are exactly r such values $p \in \mathbb{Z}_m$ that $-\frac{1}{2} < p - d\frac{m}{r} \leq \frac{1}{2}$; namely, the nearest integers to $d\frac{m}{r}$ for each $d \in \{0, 1, \ldots, r-1\}$. Therefore, the probability of seeing some is at least $\frac{2}{5}$ if $n \geq 100$. □

We have learned that an individual $p \in \mathbb{Z}_m$ such that

$$-\frac{1}{2} < p - d\frac{m}{r} \leq \frac{1}{2} \tag{4.40}$$

is observed with a probability of at least $\frac{2}{5r}$. Therefore, any corresponding $d \in \{0, 1, 2, \ldots, r-1\}$ is also obtained with a probability of at least $\frac{2}{5r}$. We should finally estimate what the probability is that for such d $\gcd(d, r) = 1$ also holds.

The probability that $\gcd(d, r) = 1$ for a given element $d \in \{0, 1, \ldots, r-1\}$ is clearly $\frac{\varphi(r)}{r}$, which can be estimated by using the following result.

Theorem 4.3.1 **([76]).** *For* $r \geq 3$,

$$\frac{r}{\varphi(r)} < e^{\gamma} \log\log r + \frac{2.50637}{\log\log r},$$

where $\gamma = 0.5772156649\ldots$ *is Euler's constant.*[1]

Lemma 4.3.2. *For* $r \geq 19$, *the probability that, for a uniformly chosen* $d \in \{0, 1, \ldots, r-1\}$, $\gcd(d, r) = 1$ *holds is at least* $\frac{1}{4\log\log n}$.

Proof. It directly follows from Theorem 4.3.1 that for $r \geq 19$,

$$\frac{r}{\varphi(r)} < 4\log\log r.$$

Therefore,

$$\frac{\varphi(r)}{r} > \frac{1}{4\log\log r} > \frac{1}{4\log\log n},$$

which was claimed. □

We combine the following facts to get Lemma (4.3.3):

- The probability that, for a randomly (and uniformly) chosen $a \in \mathbb{Z}_n$, the order $r = \mathrm{ord}_n(a)$ is even and $a^{\frac{r}{2}} \not\equiv -1 \pmod{n}$ is at least $\frac{1}{2}$ (Lemma 4.2.4).

[1] Euler's constant is defined by $\gamma = \lim_{n\to\infty}(1 + \frac{1}{2} + \frac{1}{3} + \ldots \frac{1}{n} - \log n)$. In [76] it is even shown that in the above theorem, 2.50637 can be replaced with 2.5 for all numbers r but $r = 2 \cdot 3 \cdot 5 \cdot 7 \cdot 11 \cdot 13 \cdot 17 \cdot 19 \cdot 23$.

- The probability that observing (4.31) will give a p such that $\left|p - d\frac{m}{r}\right| < \frac{1}{2}$ is at least $\frac{2}{5}$ (Lemma 4.3.1).
- The probability that $\gcd(d, r) = 1$ is at least $\frac{1}{4 \log\log n}$ (Lemma 4.3.2).

Lemma 4.3.3. *The probability that the quantum algorithm finds the order of an element of $\mathbb{Z}_n$ is at least*

$$\frac{1}{20}\frac{1}{\log\log n}.$$

Remark 4.3.1. It was already mentioned in Section 4.1.4 that many interesting quantum algorithms are based on conveying some information of interest into the coefficients of quantum superposition and then applying fast QFT. The period finding in Shor's factoring algorithm is also based on this method, as we can see: to use quantum parallelism, it prepared a superposition

$$\frac{1}{\sqrt{m}}\sum_{k=0}^{m-1} |k\rangle\,|0\rangle\,.$$

The information of interest, namely, the period, was moved into the coefficients by computing $k \mapsto a^k \pmod{n}$ to get

$$\begin{aligned}&\frac{1}{\sqrt{m}}\sum_{k=0}^{m-1} |k\rangle\,\left|a^k\right\rangle\\ &= \frac{1}{\sqrt{m}}\sum_{l=0}^{r-1}\sum_{q=0}^{s_l} |qr+l\rangle\,\left|a^l\right\rangle\,.\end{aligned}$$

In the above superposition, each $\left|a^l\right\rangle$ has

$$\sum_{q=0}^{s_l} |qr+l\rangle \tag{4.41}$$

as the left multiplier. But in (4.41), a superposition of basis vectors $|x\rangle$, $x \in \mathbb{Z}_m$, already contains information about the period in the coefficients: the coefficient of $|x\rangle$ is nonzero if and only if $x = qr + l$, so the coefficient sequence of (4.41) also has period r, and this period is to be extracted by using QFT.

Notice that the elements a^l are distinct for $l \in \{0, 1, \ldots, r-1\}$, and hence it is guaranteed that the left multipliers (4.41) do not interfere. Thus, we can say that the computation $k \mapsto a^k$ was the operative step in conveying the periodicity to the superposition's coefficients.

4.3.3 The Complexity of Shor's Factorization Algorithm

We can summarize the previous sections in the following algorithm:

Shor's quantum factorization algorithm

Input: An odd number n that has at least two distinct prime factors.

Output: A nontrivial factor of n.

1. Choose an arbitrary $a \in \{1, 2, \ldots n-1\}$.
2. Compute $d = \gcd(a, n)$ by using Euclid's algorithm. If $d > 1$, output d and stop.
3. Compute number r by using the quantum algorithm in Section 4.2.3 ($m = 2^l$ is chosen such that $n^2 \leq m < 2n^2$). If $a^r \not\equiv 1 \pmod{n}$ or r is odd or $a^{\frac{r}{2}} \equiv -1 \pmod{n}$, output "failure" and stop.
4. Compute $d_\pm = \gcd(n, a^{\frac{r}{2}} \pm 1)$ by using Euclid's algorithm. Output numbers $d_\pm$ and stop.

Step 2 can be done in time $O(\ell(n)^3)$.[2] For step 3 we first check whether a has order less than 19, which can be done in time $O(\ell(n)^3)$. Then we use Hadamard-Walsh transform in $\mathbb{Z}_m$, which can be done in time $O(\ell(m)) = O(\ell(n))$. After that, computing $a^k \pmod{n}$ can be done in time $O(\ell(m)\ell(n)^2) = O(\ell(n)^3)$. QFT in $\mathbb{Z}_m$ will be done in $O(\ell(m)^2)$ steps, and finally the computation of the convergents can be done in time $O(\ell(n)^3)$. Step 4 can also be done in time $O(\ell(n)^3)$.

The overall complexity of the above algorithm is therefore $O(\ell(n)^3)$, but the success probability is only guaranteed to be at least $\Omega(\frac{1}{\log\log n}) = \Omega(\frac{1}{\log \ell(n)})$. Thus, by running the above algorithm $O(\log(\ell(n)))$ times, we obtain a method that extracts a nontrivial factor of n in time $O(\ell(n)^3 \log \ell(n))$ with high probability.

4.4 Excercises

1. Let G be an abelian group. Find the Fourier transform of $f : G \to \mathbb{C}$ defined by

$$f(g) = \begin{cases} 1, & \text{if } g = g_i, \\ 0, & \text{otherwise.} \end{cases}$$

If f is viewed as a superposition (see connection between formulae (4.3) and (4.4)), which is the state of H corresponding to f?

---

[2] Recall that the notation $\ell(n)$ stands for the *length* of number n; that is, the number of the digits needed to represent n. This notation, of course, depends on a particular number system chosen, but in different systems (excluding the unary system) the length differs only by a multiplicative constant, and this difference can be embedded into the O-notation.

2. Let $n = n_1 n_2$, where $\gcd(n_1, n_2) = 1$. Let also

$$F : \mathbb{Z}_{n_1} \times \mathbb{Z}_{n_2} \to \mathbb{Z}_n$$

be the function given by $F((k_1, k_2)) = a_1 n_2 k_1 + a_2 n_1 k_2$, where a_1 (resp. a_2) given by the Chinese Remainder Theorem, is the multiplicative inverse of n_2 (resp. n_1) modulo n_1 (resp. n_2). Show that F is an isomorphism.
3. a) Let n and k be fixed natural numbers. Device a polynomial-time algorithm that decides whether $n = x^k$ for some integer x.
   b) Based on the above algorithm, device a polynomial-time algorithm which tells whether a given natural number n is a nontrivial power of another number.
4. Let $n = p_1^{e_1} \cdots p_k^{e_k}$ be the prime factorization of n and $a \in \mathbb{Z}_n^*$. Let $(a_1, \ldots, a_k)$ be the decomposition of a given by the Chinese Remainder Theorem and $r_i = \mathrm{ord}_{p_i^{e_i}}(a_i)$. Show that $\mathrm{ord}_n(a) = \mathrm{lcm}\{r_1, \ldots, r_k\}$.
5. a) Prove that $\left|\mathrm{e}^{ix} - 1\right|^2 = 4\sin^2 \frac{x}{2}$.
   b) Prove that

$$\left|\sum_{q=0}^{s} \mathrm{e}^{\frac{2\pi i p q r}{m}}\right|^2 = \frac{\sin^2 \frac{\pi p r (s+1)}{m}}{\sin^2 \frac{\pi p r}{m}}.$$

6. Use Taylor series to show that $\sin^2 \frac{\pi}{2}(1 + x) \geq 1 - (\frac{\pi}{2}x)^2$.

5. Finding the Hidden Subgroup

In this brief chapter we present a quantum algorithm, which can be seen as a generalization of Shor's algorithm. We can here explain, in a more structural manner, why quantum computers can speed up some computational problems. The so-called Simon's hidden subgroup problem can be stated in a general form as follows [18]:

Input: A finite abelian group G and function $\rho : G \to R$, where R is a finite set.

Promise: There exists a nontrivial subgroup $H \leq G$ such that ρ is constant and distinct on each coset of H.

Output: A generating set for H.

The function ρ is said to fulfill *Simon's promise* with respect to subgroup H. If $h \in H$, then elements g and $g + h$ belong to the same coset of H (see Section 9.1 for details), and since ρ is constant on the cosets of H, we have that $\rho(g + h) = \rho(g)$. We also say that ρ is H-periodic.

In Section 5.2 we will see that some interesting computation problems can be reduced to solving the hidden subgroup problem. In a typical example of this problem, $|G|$ is so large that an exhaustive search for the generators of H would be hopeless, but the *representation size*[1] of an individual element $g \in G$ is only $\Theta(\log |G|)$. Moreover, G can be typically described by a small number of generators. Here we will consider the description size of G, which is usually of order $(\log |G|)^k$ for some fixed k as the size of the input. It is also usually assumed that function $\rho : G \to R$ can be computed rapidly (in polynomial time) with respect to $\log |G|$.

For better illustration we will present the generalized form of Simon's quantum algorithm on $\mathbb{F}_2^m$ for solving the hidden subgroup problem. Note also that in group $\mathbb{F}_2^m$ any subgroup is a subspace as well, so instead of asking for a generating set, we could ask for a basis. By using the Gauss-Jordan elimination method one can efficiently find a basis when a generating set is known (see Exercise 4).

[1] The elements of G can be represented as binary strings, for instance.

5.1 Generalized Simon's Algorithm

In this section we will show how to solve Simon's subgroup problem on $G = \mathbb{F}_2^m$, the addivite group of m-dimensional vector space over binary field $\mathbb{F}_2 = \{0, 1\}$.

5.1.1 Preliminaries

The results represented here will also apply if $\mathbb{F}_2$ is replaced with any finite field $\mathbb{F}$, so we will, for a short moment, describe them in a more general form. More mathematical background can be found in Section 9.3.

Here we temporarily use another notation for the inner product: if $\boldsymbol{x} = (x_1, \ldots, x_m)$ and $\boldsymbol{y} = (y_1, \ldots, y_m)$ are elements of $\mathbb{F}^m$, we denote their inner product by

$$\boldsymbol{x} \cdot \boldsymbol{y} = x_1 y_1 + \ldots + x_m y_m.$$

An element $\boldsymbol{y}$ is said to be orthogonal to H if $\boldsymbol{y} \cdot \boldsymbol{h} = 0$ for each $\boldsymbol{h} \in H$. It is easy to verify that elements orthogonal to H form a subgroup (even a subspace) of $\mathbb{F}^m$, which is denoted by $H^\perp$. The importance of the following simple lemma will become clear in a moment.

Lemma 5.1.1. *For any* $\boldsymbol{y} \in \mathbb{F}^m$,

$$\sum_{\boldsymbol{h} \in H} (-1)^{\boldsymbol{h} \cdot \boldsymbol{y}} = \begin{cases} |H|, & \text{if } \boldsymbol{y} \in H^\perp, \\ 0, & \text{otherwise.} \end{cases} \tag{5.1}$$

Proof. This is analogous to proving the orthogonality of characters (Section 9.2.2). If $\boldsymbol{y} \in H^\perp$, then $\boldsymbol{h} \cdot \boldsymbol{y} = 0$ for each $\boldsymbol{h} \in H$, and the claim is obvious. If $\boldsymbol{y} \notin H^\perp$, then there exists an element $\boldsymbol{h}_1 \in H$ such that $\boldsymbol{h}_1 \cdot \boldsymbol{y} \neq 0$, and therefore

$$\begin{aligned} S = \sum_{\boldsymbol{h} \in H} (-1)^{\boldsymbol{h} \cdot \boldsymbol{y}} &= \sum_{\boldsymbol{h} \in H} (-1)^{(\boldsymbol{h} + \boldsymbol{h}_1) \cdot \boldsymbol{y}} \\ &= (-1)^{\boldsymbol{h}_1 \cdot \boldsymbol{y}} \sum_{\boldsymbol{h} \in H} (-1)^{\boldsymbol{h} \cdot \boldsymbol{y}} = (-1)^{\boldsymbol{h}_1 \cdot \boldsymbol{y}} S, \end{aligned}$$

hence $S = 0$. □

Let $H \in \mathbb{F}^m$ be a subspace having dimension d. If X is any generating subset of H, we can, by Exercise (4), efficiently find a basis of H. Any $d \times m$ matrix M over $\mathbb{F}$, whose rows are a basis of H, is called a *generator matrix* of H. The name "generator matrix" is well justified by Exercise (1). It follows from Exercises (2), (3), (4), and (5) that, once a generator matrix of H is given, we can efficiently compute a generator matrix of $H^\perp$. Moreover, since $H = (H^\perp)^\perp$ (verify), it follows that in order to find a generating set for H, it suffices to find a generating set for $H^\perp$.

For a subgroup H of $\mathbb{F}^m$, let T be some set that consists of exactly one element from each coset (see Section 9.1.2) of H. Such a set is called a *transversal* of H in $\mathbb{F}^m$ and it is clear that $|T| = [\mathbb{F}^m : H] = \frac{|\mathbb{F}^m|}{H} = \frac{2^m}{H}$. It is also easy to verify that, since $\mathbb{F}^m = H \oplus H^\perp$, the equation $|\mathbb{F}^m| = |H| \cdot |H^\perp|$ must hold.

5.1.2 The Algorithms

We will use m qubits to represent the elements of $\mathbb{F}_2^m$ and approximately $\log_2[\mathbb{F}_2^m : H] = \log_2 \frac{2^m}{|H|} = m - \log_2 |H|$ qubits for representing the values of the function $\rho : \mathbb{F}_2 \to R$. Here the description size of $G = \mathbb{F}_2^m$ is only m, and we assume that ρ is computable in polynomial time with respect to the input size.

Using only function ρ, we can find a set of generators for H. It can be shown [18] that if no other knowledge on ρ is given, i.e., ρ is given as a *blackbox* function (see Section 6.1.2), solving this problem using any classical algorithm requires exponential time, even when allowing a bounded error probability. On the other hand, we will now demonstrate that this problem can be solved in polynomial time by using a quantum circuit.

The algorithm for finding a basis of H consists of finding a basis Y of $H^\perp$, then computing the basis of H. As mentioned earlier, the latter stage can be efficiently performed by using a classical computer. The problem of finding a basis for $H^\perp$ would be easier if we knew the *dimension* of $H^\perp$ in advance, as we will see. In that case, the algorithm could be described as follows.

Algorithm A:
Finding the Basis (Dimension Known)

1. If $d = \dim H^\perp = 0$, output $\emptyset$ and stop.
2. Use the Algorithm B below to choose the set Y of d elements of $H^\perp$ uniformly.
3. Use the Gauss-Jordan elimination method (Exercise 4) to check whether Y is an independent set. If Y is independent, output Y and stop; otherwise, give "failure" as output.

In a moment, we will see that the second step of the above algorithm will give a basis of $H^\perp$ with a probability of at least $\frac{1}{4}$ (Lemma 5.1.2). Before that, we will describe how to perform that step rapidly by using a quantum circuit. It is worth mentioning that, in the above algorithm, a quantum computer is needed only to produce a set of elements of $H^\perp$.

Algorithm B:
Choosing Elements Uniformly

1. Using the Hadamard-Walsh transform on $\mathbb{F}_2^m$, prepare

$$\frac{1}{\sqrt{2^m}} \sum_{\boldsymbol{x} \in \mathbb{F}_2^m} |\boldsymbol{x}\rangle |\boldsymbol{0}\rangle . \tag{5.2}$$

2. Compute ρ to get

$$\frac{1}{\sqrt{2^m}} \sum_{\boldsymbol{x} \in \mathbb{F}_2^m} |\boldsymbol{x}\rangle |\rho(\boldsymbol{x})\rangle = \frac{1}{\sqrt{2^m}} \sum_{\boldsymbol{t} \in T} \sum_{\boldsymbol{x} \in H} |\boldsymbol{t} + \boldsymbol{x}\rangle |\rho(\boldsymbol{t})\rangle . \tag{5.3}$$

3. Use the Hadamard-Walsh transform on $\mathbb{F}_2^m$ to get

$$\begin{aligned} &\frac{1}{\sqrt{2^m}} \sum_{\boldsymbol{t} \in T} \sum_{\boldsymbol{x} \in H} \frac{1}{\sqrt{2^m}} \sum_{\boldsymbol{y} \in \mathbb{F}_2^m} (-1)^{(\boldsymbol{t}+\boldsymbol{x})\cdot \boldsymbol{y}} |\boldsymbol{y}\rangle |\rho(\boldsymbol{t})\rangle \\ &= \frac{1}{2^m} \sum_{\boldsymbol{t} \in T} \sum_{\boldsymbol{y} \in \mathbb{F}_2^m} (-1)^{\boldsymbol{t}\cdot \boldsymbol{y}} \sum_{\boldsymbol{x} \in H} (-1)^{\boldsymbol{x}\cdot \boldsymbol{y}} |\boldsymbol{y}\rangle |\rho(\boldsymbol{t})\rangle \\ &= \frac{|H|}{2^m} \sum_{\boldsymbol{t} \in T} \sum_{\boldsymbol{y} \in H^\perp} (-1)^{\boldsymbol{t}\cdot \boldsymbol{y}} |\boldsymbol{y}\rangle |\rho(\boldsymbol{t})\rangle . \end{aligned}$$

4. Make an observation to get an element $\boldsymbol{y} \in H^\perp$.

In the step 3, the equality between the the last and the second-last formulae follows from Lemma 5.1.1. It is easy to see that the probability of observing a particular $\boldsymbol{y} \in H^\perp$ is

$$P(\boldsymbol{y}) = \sum_{\boldsymbol{t} \in T} \left(\frac{|H|}{2^m} (-1)^{\boldsymbol{t}\cdot \boldsymbol{y}}\right)^2 = \frac{T\,|H|^2}{2^{2m}} = \frac{|\mathbb{F}_2^m|}{|H|} \frac{|H|^2}{|\mathbb{F}_2^m|^2} = \frac{1}{|H^\perp|},$$

so the Algorithm B works correctly: elements of $H^\perp$ are drawn uniformly. It is also clear that the above quantum algorithm runs in polynomial time.

For the second step of Algorithm A, we will now analyze the probability that when choosing d elements uniformly from a d-dimensional subspace, the chosen vectors are indpendent. The analysis is straightforward as seen in the following lemma.

Lemma 5.1.2. *Let $t \le d$ and $\boldsymbol{y}_1, \ldots, \boldsymbol{y}_t$ be randomly chosen vectors, with uniform distribution, in a d-dimensional vector space over $\mathbb{F}_2$. Then the probability that $\boldsymbol{y}_1, \ldots, \boldsymbol{y}_t$ are linearly independent is at least $\frac{1}{4}$.*

Proof. The cardinality of a d-dimensional vector space over $\mathbb{F}_2$ is 2^d. Therefore, the probability that $\{\boldsymbol{y}_1\}$ is a linearly independent set is $\frac{2^d-1}{2^d}$, since only choosing $\boldsymbol{y}_1 = \boldsymbol{0}$ makes $\{\boldsymbol{y}_1\}$ dependent. Suppose now that $\boldsymbol{y}_1, \ldots, \boldsymbol{y}_i$ have been chosen in such a way that $S = \{\boldsymbol{y}_1, \ldots, \boldsymbol{y}_{i-1}\}$ is a linearly independent set. Now S generates a subspace of dimension $i-1$. Hence, there are 2^{i-1} choices for $\boldsymbol{y}_i$ that make $S \cup \{\boldsymbol{y}_i\}$ linearly dependent. Thus, the probability that the set $\{\boldsymbol{y}_1, \ldots, \boldsymbol{y}_t\}$ is an independent set is

$$p = \frac{2^d - 2^0}{2^d} \frac{2^d - 2^1}{2^d} \cdot \ldots \cdot \frac{2^d - 2^{t-1}}{2^d} . \tag{5.4}$$

Then

$$\begin{aligned}\frac{1}{p} &= \prod_{i=0}^{t-1} \frac{2^d}{2^d - 2^i} = \prod_{i=0}^{t-1} \frac{2^{d-i}}{2^{d-i} - 1} = \prod_{i=1}^{t} \frac{2^{d-t+i}}{2^{d-t+i} - 1} \\ &\leq \prod_{i=1}^{t} \frac{2^i}{2^i - 1} = 2\prod_{i=2}^{t}\left(1 + \frac{1}{2^i - 1}\right).\end{aligned}$$

Now

$$\begin{aligned}\ln\frac{1}{2p} &\leq \sum_{i=2}^{t} \ln\left(1 + \frac{1}{2^i - 1}\right) \leq \sum_{i=2}^{t} \frac{1}{2^i - 1} \leq \sum_{i=2}^{t} \frac{1}{\frac{3}{4}\cdot 2^i} < \frac{4}{3}\sum_{i=2}^{\infty} \frac{1}{2^i} \\ &= \frac{4}{3}\cdot\frac{1}{2} = \frac{2}{3},\end{aligned}$$

so $\frac{1}{2p} < e^{\frac{2}{3}} < 2$, hence $p > \frac{1}{4}$. □

Remark 5.1.1. Notice that if we build another algorithm (call it A′) which repeats Algorithm A whenever the output is "failure", then A′ cannot give us a wrong answer, but we cannot find a priori upper bound for the number of repeats it has to make. Instead, we can say that *on average*, the number of repeats is at most four. Such an algorithm is called a *Las Vegas algorithm*, cf. Section 3.1.2. An interesting method for finding the basis in polynomial time with *certainty* is described in [18].

Remark 5.1.2. Algorithm B for choosing an element in $H^\perp$ resembles the period-finding algorithm of Section 4.2.3. The first step was to prepare a quantum superposition of all the elements of $\mathbb{F}_2^m$. Then, by utilizing quantum parallelism, function ρ was computed simultaneously on all elements of $\mathbb{F}_2^m$. This was the operative step to convey information of interest to the superposition coefficients, and this information was extracted by using QFT. Notice that since ρ is different on distinct cosets, all the vectors $|\rho(\boldsymbol{t})\rangle$ are orthogonal and, therefore, the states

$$\sum_{\boldsymbol{y}\in H^\perp} (-1)^{\boldsymbol{t}\cdot\boldsymbol{y}} |\boldsymbol{y}\rangle\, |\rho(\boldsymbol{t})\rangle$$

with different values of $\rho(\boldsymbol{t})$ do not interfere.

To conclude this section, we describe how to find the basis of $H^\perp$, when $d = \dim H^\perp$ is not known in advance. Because of Algorithm A, this is done if we can describe an algorithm which can find out the dimension of $H^\perp$.

The dimension of $H^\perp$ can be found by utilizing Algorithm A (which uses Algorithm B as a subprocedure). By Lemma 5.1.2 we see that, if we choose uniformly $t \leq d$ elements from a subspace of dimension d, then the probability that the chosen elements are independent is at least $\frac{1}{4}$. We utilize this as follows: if $D = \dim H^\perp$, the (unknown) dimension of $H^\perp$, we just *guess* that the dimension of $H^\perp$ is d ($0 \leq d \leq m$), and run Algorithm A (with the d we guessed). If $d \leq D$, we obtain, according to Lemma 5.1.2, a set of d linearly independent vectors with a probability of at least $\frac{1}{4}$. Once we get such a set of vectors, we can tell, *with certainty*, that the guess $d \leq D$ was correct (the number of linearly independent vectors that a vector space can contain is at most the dimension of the space).

On the other hand, if $d > D$, the set of d vectors we obtain can never be linearly independent. After repeating Algorithm A a number of times, we can be convinced that our guess $d > D$ was incorrect. Anyway, Algorithm A can always be used to decide whether $d \leq D$ holds with a correctness probability of at least $\frac{1}{4}$.

We will now describe how to find D, the dimension of $H^\perp$. To make the description easier, we will regard all other elements (D, H, and ρ) fixed, but only the interval I where the dimension may be found, is mentioned as an input.

For an interval $I = \{k, k+1, \ldots, k+r\}$ we use notation $M(I) = \lfloor \frac{r}{2} \rfloor$ for the "midpoint" of I, $B(I) = \{k, k+1, \ldots, M(I)+1\}$ for the "initial part" and $T(I) = \{M(I), \ldots, k+r\}$ for the "final part" of I. In the following algorithm, D always stands for the true (unknown) dimension of $H^\perp$ and initially, the input of the following algorithm is the interval $I = \{0, 1, \ldots, m\}$.

Algorithm C:
Determining the Dimension

1. If $|I| = 1$, output the unique element of I and stop.
2. If $|I| > 1$, run Algorithm A to decide if $D \leq M(I)$. If the answer is "yes", apply this algorithm with input $B(I)$; otherwise, apply this algorithm with input $T(I)$.

Algorithm C thus finds the dimension D by recursively cutting the interval $\{0, 1, \ldots, m\}$ into two approximately equally long parts, and using Algorithm A to decide which one is the part containing D. It is therefore clear that Algorithm C stops after $\Theta(\log_2 m)$ recursive calls. Since running Algorithm C, we have to make only $k = \Theta(\log_2 m)$ decisions based on Algorithm A, which has a correctness probability of at most $\frac{1}{4}$, we have that Algorithm C works with a correctness probability of $p \geq (\frac{1}{4})^k = \Theta(\frac{1}{4}^{\log_2 m}) = \Theta(\frac{1}{m^2})$. Therefore, repeating Algorithm C $O(m^2)$ times we get an algorithm which gives the dimension D with a nonvanishing correctness probability.

It is left as an exercise to analyze the running time of Algorithms A, B, and C to conclude that the generators of a hidden subgroup can be found in polynomial time provided that ρ can be computed in polynomial time.

Remark 5.1.3. It is an open problem whether the hidden subgroup problem can be solved, by using a quantum computer, in polynomial time for *non-abelian groups* as well. For progress in this direction, see [36] and the references therein. The hidden subgroup problem for non-abelian groups is of special interest, since the *graph isomorphism problem* can be reduced to it.

5.2 Examples

We will now demonstrate how the hidden subgroup problem can be applied to computational problems. These examples are due to [59].

5.2.1 Finding the Order

This problem lies at the very heart of Shor's factorization algorithm: Let n be a large integer and a another integer such that $\gcd(a, n) = 1$. Denote $r = \text{ord}_n(a)$, i.e., r is the least positive integer that satisfies $a^r \equiv 1 \pmod{n}$. In this problem, we have a group $\mathbb{Z}$, and the hidden subgroup is $r\mathbb{Z}$, whose generator r should be found.

The function $\rho(x) = a^x + n\mathbb{Z}$ satisfies Simon's promise, since

$$\begin{aligned} & \rho(x) = \rho(y) \\ \iff & a^x + n\mathbb{Z} = a^y + n\mathbb{Z} \\ \iff & n \mid (a^x - a^y) \\ \iff & r \mid (x - y) \\ \iff & x + r\mathbb{Z} = y + r\mathbb{Z}. \end{aligned}$$

Because $\mathbb{Z}$ is an infinite group, we cannot directly solve this problem by using the algorithm of the previous section, but using finite groups $\mathbb{Z}_{2^l}$ instead of $\mathbb{Z}$ already gives us approximations that are good enough, as shown in the previous chapter.

5.2.2 Discrete Logarithm

If F is a cyclic group of order q, then there is a generator $g \in F$ such that $F = \{1 = g^0, g^1, g^2, \ldots, g^{q-1}\}$ (see Section 9.1 for notions of group theory). Hence, if we are given a generator g of F and another element $a \in F$, then a can be uniquely expressed as $a = g^r$, where $r \in \{0, 1, 2, \ldots, q-1\}$. The *discrete logarithm problem* is to find r.

Since $g^q = 1$ (and q is the least positive number with this property), we have that $g^{r_1} = g^{r_2}$ if and only if $r_1 \equiv r_2 \pmod{q}$. Hence, instead of regarding the exponent of g as an integer, we can regard it as an element of $\mathbb{Z}_q$ as well. In fact, mapping $\mathbb{Z}_q \to F$, $r + q\mathbb{Z} \mapsto g^r$ is an isomorphism.

We can reduce the discrete logarithm to finding the hidden subgroup as follows:

If $a = g^r \in F$, let $G = \mathbb{Z}_q \times \mathbb{Z}_q$. The hidden subgroup $H = \{(a, ra) \mid a \in \mathbb{Z}_q\}$ is generated by element $(1, r)$. That function $\rho : \mathbb{Z}_q \times \mathbb{Z}_q \to G$ defined as

$$\rho(x, y) = g^x a^{-y}$$

satisfies Simon's promise, can be verified easily:

$$\begin{aligned} & \rho(x_1, y_1) = \rho(x_2, y_2) \\ \iff & g^{x_1} a^{-y_1} = g^{x_2} a^{-y_2} \\ \iff & g^{x_1 - r y_1} = g^{x_2 - r y_2} \end{aligned}$$

But the last equation is true if and only if $x_1 - ry_1 = x_2 - ry_2$ (when regarded as elements of $\mathbb{Z}_q$), which, in turn, is true if and only if $x_1 - x_2 = r(y_1 - y_2)$. This is equivalent to condition $(x_1, y_1) + H = (x_2, y_2) + H$.

On the other hand, there may be other generators of H than $(1, r)$, but using a number-theoretic argumentation similar to that used in Section 4.3.2, we can see that the discrete logarithm can be found with a nonvanishing probability in polynomial time.

Remark 5.2.1. The reliability of U.S. Digital Signature Algorithm is based on the assumption that the discrete logarithm problem on *finite fields* remains intractable on classical computers [60].

5.2.3 Simon's Original Problem

Simon [84] originally formulated his problem as follows:

Input: An integer $m \geq 1$ and a function $\rho : \mathbb{F}_2^m \to R$, where R is a finite set.

Promise: There exists a nonzero element $\boldsymbol{s} \in \mathbb{F}_2^m$ such that for all $\boldsymbol{x}, \boldsymbol{y} \in \mathbb{F}_2^m$ $\rho(\boldsymbol{x}) = \rho(\boldsymbol{y})$ if and only if $\boldsymbol{x} = \boldsymbol{y}$ or $\boldsymbol{x} = \boldsymbol{y} + \boldsymbol{s}$.

Output: Element $\boldsymbol{s}$.

It is plain to see that choosing G as $\mathbb{F}_2^m$ and $H = \{\boldsymbol{0}, \boldsymbol{s}\}$ as the subspace generated by $\boldsymbol{s}$ makes this problem a special case of a more general problem presented at the beginning of this chapter.

Simon's problem is of historical interest, since it was the first one providing an example of a problem that can be solved, with nonvanishing probability, in polynomial time by using a quantum computer, but which requires exponential time in any classical algorithm if ρ is considered a blackbox function (see Section 6.1.2).

5.3 Exercises

1. Show that if M is a generator matrix of H, then
$$H = \{\boldsymbol{y}M \mid \boldsymbol{y} \in \mathbb{F}^d\}.$$
2. Show that if M is a generator matrix of H and N is a $d' \times m$ matrix such that $MN^{\perp} = 0$, then the subspace generated by the rows of N is orthogonal to H.
3. The *elementary row operations* on a matrix over field $\mathbb{F}$ are:

a) Swapping two rows, $\boldsymbol{x}_i \leftrightarrow \boldsymbol{x}_j$.
b) Multiplying a row with a nonzero element of $\mathbb{F}$, $\boldsymbol{x}_i \mapsto c\boldsymbol{x}_i$, $c \neq 0$.
c) Adding to a row another row multiplied by any element of $\mathbb{F}$, $\boldsymbol{x}_i \mapsto \boldsymbol{x}_i + c\boldsymbol{x}_j$.

Show that if M' is obtained from M by using the elementary row operations, then the rows of M and M' generate exactly the same subspace.

4. A matrix is in a *reduced row-echelon form* if:
   a) For any row containing a nonzero entry, the first such is 1. The first nonzero entry in a row is called the *pivot*.
   b) The rows containing only zero entries, if there are such rows, are the last ones.
   c) The pivot on each row is on the right-hand side of the pivot of the row above.
   d) Each *column* containing a pivot of a row has zeros everywhere else.

   If only conditions a), b), and c) are satisfied, we say that the matrix is in *row-echelon form*.
   Prove the following *Gauss-Jordan elimination theorem*. Each $k \times m$ matrix can be transformed into a reduced row-echelon form by using $O(k^3 m)$ field operations. Moreover, the nonzero rows of a row-echelon form matrix are linearly independent.
5. Prove that if $M = (I_d \mid M')$ is a generator $(d \times (m-d))$ matrix of H, then $N = (-M^T \mid I_{m-d})$ is a generator matrix of H^T.

6. Grover's Search Algorithm

6.1 Search Problems

Let us consider a children's game called *hiding the key*: one player hides the key in the house, and the others try to find it. The one who has hidden the key is permanently advising the others by using phrases like "freezing", "cold", "warm", and "hot", depending on how close the seekers are to the hiding place. Without this advice, the game would obviously last much longer. Or, can you develop a strategy for finding the key without searching through the entire house?

6.1.1 Satisfiability Problem

There are many problems in computer science that closely resemble searching for a tiny key in a large house. We will shortly discuss the problem of finding a solution for an **NP**-complete 3-satisfiability problem: we are given a propositional expression in a conjunctive normal form; each clause is a disjunction of three *literals* (a Boolean variable or a negation of a Boolean variable). In the original form of the problem, the task is to find a satisfying truth assignment, if any such exists. Let us then imagine an advisor who always has enough power to tell at once whether or not a given Boolean expression has a satisfying truth assignment. If such an advisor were provided, finding a satisfying assignment is no longer a difficult problem: we could just substitute 1 for some variable and ask the advisor whether or not the resulting Boolean expression with less variables had a demanded assignment. If our choice was incorrect, we would flip the substituted value and proceed on recursively.

Unfortunately, the advisor's problem to tell whether a satisfying valuation exists is an **NP**-complete problem, and in light of our present knowledge, it is very unlikely that there would be a fast solution to this problem in the real world. But let us continue our thinking experiment by assuming that somebody, let us call him *a verifier*, knows a satisfying assignment for a Boolean expression but is not willing to tell it to us. Quite surprisingly, there exist so-called *zero-knowledge protocols* (see [79] for more details), which the verifier can use to guarantee that he really knows a satisfying valuation *without revealing even a single bit of his knowledge.* Thus, it is possible that we are quite sure that a satisfying truth assignment exists, yet we do not have a clue what it might be! The obvious strategy to find a satisfying assignment is to search through all of the possible assignments. But if there are n Boolean variables in the expression, there are 2^n assignments, and because we do not have a supernatural advisor, our task seems quite hopeless for large n, at least in the general case. In fact, no faster method than an exhaustive search is known in the general case.

Remark 6.1.1. The most effective classical procedure for solving generic **NP**-complete problems seems to be that one described by Uwe Schönig [83].

6.1.2 Probabilistic Search

In what follows, we will slightly generalize our search problem. Instead of seeking a satisfying assignment for a Boolean expression, we will generally talk about functions

$$f : \mathbb{F}_2^n \to \mathbb{F}_2,$$

and our search problem is to find a $\boldsymbol{x} \in \mathbb{F}_2^n$ such that $f(\boldsymbol{x}) = 1$ (if any such $\boldsymbol{x}$ exists).

Notice that, with the assumption that f is computable in polynomial time, this model is enough to represent **NP** problems because it includes the **NP**-complete 3-satisfiability problem. This model is also assigned [41] to an *unordered database search*, where we have to find a specific item in a huge database. Here the database consists of 2^n items, and f is the function giving a value of 1 to a required item and 0 to the rest, hereby telling us whether the item under investigation was the required one.

Another simplification, a huge one this time, is to assume that f is a so-called *blackbox* function, i.e., we do not know how to compute f, but we can query f, and the value is returned to us instantly in one computational step. With that assumption, we will demonstate in the next chapter how to derive a lower bound for the number of queries to f to find an item $\boldsymbol{x}$ such that $f(\boldsymbol{x}) = 1$. In the next chapter, we will, in fact, present a general strategy for finding lower bounds for the number of queries concerning other goals, too.

Now we will study *probabilistic search*, i.e., we will omit the requirement that the search stragegy will *always* give such a value $\boldsymbol{x}$ that $f(\boldsymbol{x}) = 1$ but will do this only with a *nonvanishing* probability. This means that the search algorithm will give the required $\boldsymbol{x}$ with at least some constant probability $0 < p \leq 1$ for any n. This can be seen as a natural generalization of the search which with certainty returns the required item.

Remark 6.1.2. Provided that the probabilistic search strategy is rapid, we use very standard argumentation to say that we can find $\boldsymbol{x}$ rapidly *in practice*: the probability that, after m attempts we *have not* found $\boldsymbol{x}$, is at most $(1-p)^m \leq e^{-pm}$, which is smaller than any given $\epsilon > 0$, when $m > -\frac{\log \epsilon}{p}$. Thus, we can reduce the error probability p to any other positive constant ε just by repeating the original search a constant number of times.

It is easy to see that any classical deterministic search algorithm which always gives an $\boldsymbol{x}$ such that $f(\boldsymbol{x}) = 1$ (if such $\boldsymbol{x}$ exists) will require $N = 2^n$ queries to f: if some algorithm makes less than N queries, we can modify f so that the answer is no longer correct.

Remark 6.1.3. In the above argumentation, it is, of course, essential that we handle only blackbox functions. If we are, instead, given an algorithm for computing f, it is very difficult to say how easily this algorithm itself will give information about $\boldsymbol{x}$ to be found. In fact, the question about the computational work required to recover $\boldsymbol{x}$ from the algorithm for f lies at the very heart of the difficult open problem whether $\mathbf{P} \neq \mathbf{NP}$.

We cannot do much better than a deterministic search by allowing randomization: fix $\boldsymbol{y} \in \mathbb{F}_2^n$, and consider a blackbox function $f_{\boldsymbol{y}}$ defined by

$$f_{\boldsymbol{y}}(\boldsymbol{x}) = \begin{cases} 1, \text{ if } \boldsymbol{x} = \boldsymbol{y}, \\ 0, \text{ otherwise.} \end{cases} \tag{6.1}$$

If we draw disjoint elements $\boldsymbol{x}_1, \ldots, \boldsymbol{x}_k \in \mathbb{F}_2^n$ with uniform distribution, the probability of finding $\boldsymbol{y}$ is $\frac{k}{N}$, so we would need at least pN queries to find $\boldsymbol{y}$ with a probability of at least p. Using nonuniform distributions will not offer any relief:

Lemma 6.1.1. *Let $N = 2^n$ and f be a blackbox function. Assume that A_f is a probabilistic algorithm that makes queries to f and returns an element $\boldsymbol{x} \in \mathbb{F}_2^n$. If, for any nonconstant f, the probability that $f(\boldsymbol{x}) = 1$ is at least $p > 0$, then there is f such that A_f makes at least $pN - 1$ queries.*

Proof. Let $f_{\boldsymbol{y}}$ be as in (6.1) and $P_{\boldsymbol{y}}(k)$ be the probability that $A_{f_{\boldsymbol{y}}}$ returns $\boldsymbol{y}$ using k queries. By assumption $P_{\boldsymbol{y}}(k) \geq p$, and we will demonstrate that there is some $\boldsymbol{y} \in \mathbb{F}_2^n$ such that $P_{\boldsymbol{y}}(k) \leq \frac{k+1}{N}$.

First, by using induction, we show that

$$\sum_{\boldsymbol{y} \in \mathbb{F}_2^n} P_{\boldsymbol{y}}(k) \leq k + 1.$$

If $k = 0$, then $A_{f_{\boldsymbol{y}}}$ gives any $\boldsymbol{x} \in \mathbb{F}_2^n$ with some probability $p_{\boldsymbol{x}}$, and thus

$$\sum_{\boldsymbol{y} \in \mathbb{F}_2^n} P_{\boldsymbol{y}}(0) = \sum_{\boldsymbol{y} \in \mathbb{F}_2^n} p_{\boldsymbol{y}} = 1.$$

Assume, then, that

$$\sum_{\boldsymbol{y} \in \mathbb{F}_2^n} P_{\boldsymbol{y}}(k-1) \leq k.$$

On the kth query A_{f_y} queries $f(\boldsymbol{y})$ with some probability $q_{\boldsymbol{y}}$, and therefore $P_{\boldsymbol{y}}(k) \leq P_{\boldsymbol{y}}(k-1) + q_{\boldsymbol{y}}$. Thus,

$$\sum_{\boldsymbol{y}\in\mathbb{F}_2^n} P_{\boldsymbol{y}}(k) \leq \sum_{\boldsymbol{y}\in\mathbb{F}_2^n} P_{\boldsymbol{y}}(k-1) + \sum_{\boldsymbol{y}\in\mathbb{F}_2^n} q_{\boldsymbol{y}} \leq k+1.$$

Because there are $N = 2^n$ different choices for $\boldsymbol{y}$, there must exist one with

$$P_{\boldsymbol{y}}(k) \leq \frac{k+1}{N}.$$

It follows that $\frac{k+1}{N} \geq P_{\boldsymbol{y}}(k) \geq p$, so $k \geq pN - 1$. □

6.1.3 Quantum Search with One Query

Let us continue studying blackbox functions $f : \mathbb{F}_2^n \to \mathbb{F}_2$. The natural question arising now is whether we can devise a faster search method on a quantum computer using quantum parallelism for querying many values of a blackbox function simultaneously.

The very first thing we need to fix is the notion of a *quantum blackbox function.* The notion we follow here is widely accepted: let $\boldsymbol{x} \in \mathbb{F}_2^n$. In order to make a blackbox function query $f(\boldsymbol{x})$ on a quantum computer, we will utilize a *source register* $|\boldsymbol{x}\rangle$ (n qubits) and a *target qubit* $|b\rangle$. A *query operator* Q_f is a linear mapping defined by

$$Q_f \,|\boldsymbol{x}\rangle\,|b\rangle = |\boldsymbol{x}\rangle\,|b \oplus f(\boldsymbol{x})\rangle\,, \tag{6.2}$$

where $\oplus$ means addition modulo 2 or, in other words, exclusive or -operation.

We can now easily see that (6.2) defines a unitary mapping. In fact, vector set

$$B = \{|\boldsymbol{x}\rangle\,|b\rangle \mid \boldsymbol{x} \in \mathbb{F}_2^n, b \in \mathbb{F}_2\}$$

spans a 2^{n+1}-dimensional Hilbert space. Mapping Q_f operates as a permutation on B, and it is a well-known fact that any such mapping is unitary. Moreover, since

$$Q_f Q_f \,|\boldsymbol{x}\rangle\,|b\rangle = |\boldsymbol{x}\rangle\,|b \oplus f(\boldsymbol{x}) \oplus f(\boldsymbol{x})\rangle = |\boldsymbol{x}\rangle\,|b\rangle\,,$$

the inverse of Q_f is Q_f itself.

Let us now fix $\boldsymbol{y} \in \mathbb{F}_2^n$ and try to devise a quantum search algorithm for function $f_{\boldsymbol{y}}$ defined in (6.1). The very first idea is to generate a state

$$\frac{1}{\sqrt{2^n}} \sum_{\boldsymbol{x}\in\mathbb{F}_2^n} |\boldsymbol{x}\rangle\,|0\rangle \tag{6.3}$$

beginning with $|\mathbf{0}\rangle\,|0\rangle$ and using Hadamard-Walsh transform H_n (cf. Section 4.1.2). After that, one could make a single query Q_{f_y} to obtain state

$$\frac{1}{\sqrt{2^n}} \sum_{\boldsymbol{x}\in\mathbb{F}_2^n} |\boldsymbol{x}\rangle\, |0 \oplus f_{\boldsymbol{y}}(\boldsymbol{x})\rangle = \frac{1}{\sqrt{2^n}} \sum_{\boldsymbol{x}\in\mathbb{F}_2^n} |\boldsymbol{x}\rangle\, |f_{\boldsymbol{y}}(\boldsymbol{x})\rangle\,. \tag{6.4}$$

But if we now observe the last qubit of state (6.4), we would only see 1 (and after that, observing the first register, get the required $\boldsymbol{y}$) with a probability of $\frac{1}{2^n}$, so we would not gain any advantage over just guessing $\boldsymbol{y}$.

But quantum search *can* improve the probabilistic search, as we will now demonstrate. Having state (6.3), we could flip the target bit (to $|1\rangle$), and then apply the Hadamard transform to that bit to get a state

$$\frac{1}{\sqrt{2^n}} \sum_{\boldsymbol{x}\in\mathbb{F}_2^n} |\boldsymbol{x}\rangle\, \frac{1}{\sqrt{2}}(|0\rangle - |1\rangle) = \frac{1}{\sqrt{2^{n+1}}}\Big(\sum_{\boldsymbol{x}\in\mathbb{F}_2^n} |\boldsymbol{x}\rangle\, |0\rangle - \sum_{\boldsymbol{x}\in\mathbb{F}_2^n} |\boldsymbol{x}\rangle\, |1\rangle \Big). \tag{6.5}$$

If $\boldsymbol{x} \neq \boldsymbol{y}$, then $Q_{f_{\boldsymbol{y}}} |\boldsymbol{x}\rangle\, |0\rangle = |\boldsymbol{x}\rangle\, |0\rangle$ and $Q_{f_{\boldsymbol{y}}} |\boldsymbol{x}\rangle\, |1\rangle = |\boldsymbol{x}\rangle\, |1\rangle$, but $Q_{f_{\boldsymbol{y}}} |\boldsymbol{y}\rangle\, |0\rangle = |\boldsymbol{y}\rangle\, |1\rangle$ and $Q_{f_{\boldsymbol{y}}} |\boldsymbol{y}\rangle\, |1\rangle = |\boldsymbol{y}\rangle\, |0\rangle$; so, querying $f_{\boldsymbol{y}}$ by the query operator $Q_{f_{\boldsymbol{y}}}$ on state (6.5) would give us a state

$$\begin{aligned} &\frac{1}{\sqrt{2^{n+1}}}\Big(\sum_{\boldsymbol{x}\neq\boldsymbol{y}} |\boldsymbol{x}\rangle\, |0\rangle + |\boldsymbol{y}\rangle\, |1\rangle - \sum_{\boldsymbol{x}\neq\boldsymbol{y}} |\boldsymbol{x}\rangle\, |1\rangle - |\boldsymbol{y}\rangle\, |0\rangle \Big) \\ &= \frac{1}{\sqrt{2^{n+1}}}\Big(\sum_{\boldsymbol{x}\neq\boldsymbol{y}} |\boldsymbol{x}\rangle\, (|0\rangle - |1\rangle) + |\boldsymbol{y}\rangle\, (|1\rangle - |0\rangle) \Big) \\ &= \frac{1}{\sqrt{2^n}} \sum_{\boldsymbol{x}\in\mathbb{F}_2^n} (-1)^{f_{\boldsymbol{y}}(\boldsymbol{x})}\, |\boldsymbol{x}\rangle\, \frac{1}{\sqrt{2}}(|0\rangle - |1\rangle). \end{aligned} \tag{6.6}$$

Notice that preparing the target bit in superposition $\frac{1}{\sqrt{2}}(|0\rangle - |1\rangle)$ before applying the query operator was used to encode the value $f_{\boldsymbol{y}}(\boldsymbol{x})$ in the sign $(-1)^{f_{\boldsymbol{y}}(\boldsymbol{x})}$. This is a very basic technique in quantum computation. In the continuation we will not need the target bit anymore, and instead of (6.6) we will write only

$$\frac{1}{\sqrt{2^n}} \sum_{\boldsymbol{x}\in\mathbb{F}_2^n} (-1)^{f_{\boldsymbol{y}}(\boldsymbol{x})}\, |\boldsymbol{x}\rangle\,. \tag{6.7}$$

So far we have seen that applying the query operator only once in a quantum superposition (6.4), we get state (6.7) ignoring the target qubit. We continue as follows: we write (6.7) in form

$$\frac{1}{\sqrt{2^n}}\Big(\sum_{\boldsymbol{x}\in\mathbb{F}_2^n} |\boldsymbol{x}\rangle - 2\, |\boldsymbol{y}\rangle \Big),$$

and apply Hadamard H_n transform to get

$$\begin{aligned} &|\boldsymbol{0}\rangle - \frac{2}{2^n} \sum_{\boldsymbol{x}\in\mathbb{F}_2^n} (-1)^{\boldsymbol{x}\cdot\boldsymbol{y}}\, |\boldsymbol{x}\rangle \\ &= \Big(1 - \frac{2}{2^n}\Big)\, |\boldsymbol{0}\rangle - \frac{2}{2^n} \sum_{\boldsymbol{x}\neq\boldsymbol{0}} (-1)^{\boldsymbol{x}\cdot\boldsymbol{y}}\, |\boldsymbol{x}\rangle\,. \end{aligned} \tag{6.8}$$

In superposition (6.8), we separate $|\mathbf{0}\rangle$ from all other states by designing function $f_{\mathbf{0}} : \mathbb{F}_2^n \to \mathbb{F}_2$, which gets a value of 1 for $\mathbf{0}$ and 0 for all other $\boldsymbol{x} \in \mathbb{F}_2^n$. Such a function is possible to construct by using $O(n)$ Boolean gates and, through the considerations of section 3.2.3, it is also possible to construct $f_{\mathbf{0}}$ by using $O(n)$ quantum gates, using several ancilla qubits. We will store the value of $f_{\mathbf{0}}$ in an additional qubit to obtain state

$$\left(1 - \frac{2}{2^n}\right) |\mathbf{0}\rangle |1\rangle - \frac{2}{2^n} \sum_{\boldsymbol{x} \neq \mathbf{0}} (-1)^{\boldsymbol{x}\cdot\boldsymbol{y}} |\boldsymbol{x}\rangle |0\rangle . \tag{6.9}$$

Now, observation of the last qubit of (6.9) will result in

$$|\mathbf{0}\rangle |1\rangle \tag{6.10}$$

with a probability of $(1 - \frac{2}{2^n})^2 = 1 - \frac{4}{2^n} + \frac{4}{2^{2n}}$, and in

$$\begin{aligned} &\frac{1}{\sqrt{2^n - 1}} \sum_{\boldsymbol{x} \neq \mathbf{0}} (-1)^{\boldsymbol{x}\cdot\boldsymbol{y}} |\boldsymbol{x}\rangle |0\rangle \\ &= \frac{1}{\sqrt{2^n - 1}} \Big(\sum_{\boldsymbol{x} \in \mathbb{F}_2^n} (-1)^{\boldsymbol{x}\cdot\boldsymbol{y}} |\boldsymbol{x}\rangle - |\mathbf{0}\rangle \Big) |0\rangle \end{aligned} \tag{6.11}$$

with a probability of $\frac{4}{2^n} - \frac{4}{2^{2n}}$.

Applying Hadamard transform H_n once more to states (6.10) and (6.11) will yield (ignoring the last qubit)

$$\frac{1}{\sqrt{2^n}} \sum_{\boldsymbol{z} \in \mathbb{F}_2^n} |\boldsymbol{z}\rangle \tag{6.12}$$

and

$$\begin{aligned} &\frac{1}{\sqrt{2^n(2^n - 1)}} \Big(\sum_{\boldsymbol{x} \in \mathbb{F}_2^n} (-1)^{\boldsymbol{x}\cdot\boldsymbol{y}} \sum_{\boldsymbol{z} \in \mathbb{F}_2^n} (-1)^{\boldsymbol{x}\cdot\boldsymbol{z}} |\boldsymbol{z}\rangle - \sum_{\boldsymbol{z} \in \mathbb{F}_2^n} |\boldsymbol{z}\rangle \Big) \\ &= \frac{1}{\sqrt{2^n(2^n - 1)}} \Big((2^n - 1) |\boldsymbol{y}\rangle + \sum_{\boldsymbol{z} \neq \boldsymbol{y}} |\boldsymbol{z}\rangle \Big), \end{aligned} \tag{6.13}$$

respectively.

Finally, observing (6.12) and (6.13) will give us $\boldsymbol{y}$ with probabilities $\frac{1}{2^n}$ and $\frac{2^n-1}{2^n}$ respectively. Keeping in mind the probabilities to see states (6.10) and (6.11), we find that the total probability of observing $\boldsymbol{y}$ is

$$\left(1 - \frac{2}{2^n}\right)^2 \frac{1}{2^n} + \left(1 - \left(1 - \frac{2}{2^n}\right)^2\right) \frac{2^n - 1}{2^n} = \frac{1}{2^n}\left(5 - \frac{4}{2^n}\right) \approx \frac{5}{2^n},$$

which is approximately 2.5 times better than $\frac{2}{2^n}$, the best we could get by a randomized algorithm that queries $f_{\boldsymbol{y}}$ only once (see the proof of Lemma 6.1.1).

Soon we will see that, by using a quantum circuit, it is still possible to improve this probability. The next section is devoted to Lov Grover's ingenious idea of using quantum operations to amplify the probability for finding the required element.

6.2 Grover's Amplification Method

We will still use blackbox function $f_{\boldsymbol{y}} : \mathbb{F}_2^n \to \mathbb{F}_2$ from the previous section as an example. The task is to find $\boldsymbol{y}$ by querying the $f_{\boldsymbol{y}}$, and we will follow the method of Lov Grover to demonstrate that using a quantum circuit, we can find $\boldsymbol{y}$ with nonvanishing probability by using $O(\sqrt{2^n})$ queries. We will also show how to generalize this method for all blackbox functions.

6.2.1 Quantum Operators for Grover's Search Algorithm

A query operator $Q_{f_{\boldsymbol{y}}}$ which is used to call for values of $f_{\boldsymbol{y}}$ uses n qubits for the source register and 1 for the target. We will also need a quantum operator R_n defined on n qubits and operating as $R_n \left|\mathbf{0}\right\rangle = -\left|\mathbf{0}\right\rangle$ and $R_n \left|\boldsymbol{x}\right\rangle = \left|\boldsymbol{x}\right\rangle$, if $\boldsymbol{x} \neq \mathbf{0}$. If we index the rows and columns of a matrix by elements of $\mathbb{F}_2^n$ (ordered as binary numbers), we can easily express R_n as a $2^n \times 2^n$ matrix,

$$R_n = \begin{pmatrix} -1\,0\,0\ldots 0 \\ 0\,1\,0\ldots 0 \\ 0\,0\,1\ldots 0 \\ \vdots\;\vdots\;\vdots\;\ddots\;\vdots \\ 0\,0\,0\ldots 1 \end{pmatrix}. \tag{6.14}$$

But we require that the operations on a quantum circuit should be *local*, i.e., there should be a fixed upper bound on number of qubits on which each gate operates. Let us, therefore, pay some attention on how to decompose (6.14) into local operations.

The decomposition can obviously be made by making use of function $f_{\mathbf{0}} : \mathbb{F}_2^n \to \mathbb{F}_2$, which is 1 at $\mathbf{0}$ and 0 everywhere else. As discussed in the earlier section, a quantum circuit F_n for this function can be constructed by using $O(n)$ simple quantum gates (which operate on at most three qubits) and some, say k_n, ancilla qubits. To summarize: we can obtain a quantum circuit F_n on $n + k_n$ qubits operating as

$$F_n \left|\boldsymbol{x}\right\rangle \left|0\right\rangle \left|\mathbf{0}\right\rangle = \left|\boldsymbol{x}\right\rangle \left|f_{\mathbf{0}}(\boldsymbol{x})\right\rangle \left|\boldsymbol{a}_{\boldsymbol{x}}\right\rangle . \tag{6.15}$$

Using this, we can proceed as follows: we begin with

$$\left|\boldsymbol{x}\right\rangle \left|1\right\rangle \left|\mathbf{0}\right\rangle ,$$

and then operate with H_2 on the target bit to get

$$|\boldsymbol{x}\rangle \frac{1}{\sqrt{2}}(|0\rangle - |1\rangle) |\mathbf{0}\rangle .$$

After this, we call F_n to get

$$|\boldsymbol{x}\rangle (-1)^{f_O(\boldsymbol{x})} \frac{1}{\sqrt{2}}(|0\rangle - |1\rangle) |\boldsymbol{a}_{\boldsymbol{x}}\rangle .$$

Using first the *reverse circuit* F_n^{-1} and then H_2 on the target qubit will give us

$$(-1)^{f_O(\boldsymbol{x})} |\boldsymbol{x}\rangle |1\rangle |\mathbf{0}\rangle .$$

Composing all of this together, we get an operation

$$|\boldsymbol{x}\rangle |1\rangle |\mathbf{0}\rangle \mapsto (-1)^{f_O(\boldsymbol{x})} |\boldsymbol{x}\rangle |1\rangle |\mathbf{0}\rangle ,$$

which is the required operation R_n with some ancilla qubits. As usual, we will not write down the ancilla explicitly.

The other tool needed for a Grover search is to encode the value $f(\boldsymbol{x})$ in the sign - but this can be easily achieved by preparing the target bit in a superposition:

$$Q_f |\boldsymbol{x}\rangle \frac{1}{\sqrt{2}}(|0\rangle - |1\rangle) = (-1)^{f(\boldsymbol{x})} |\boldsymbol{x}\rangle \frac{1}{\sqrt{2}}(|0\rangle - |1\rangle). \tag{6.16}$$

Instead of (6.16), we will introduce notation

$$V_f |\boldsymbol{x}\rangle = (-1)^{f(\boldsymbol{x})} |\boldsymbol{x}\rangle$$

for the *modified query operator*, again omitting the ancilla qubit.

6.2.2 Amplitude Amplification

Grover's method for finding an element $\boldsymbol{y} \in \mathbb{F}_2^n$ that satisfies $f(\boldsymbol{y}) = 1$ – we will call such a $\boldsymbol{y}$ a *solution* hereafter – can be viewed as iterative amplitude amplification. The basic element is quantum operator

$$G_n = -H_n R_n H_n V_f \tag{6.17}$$

working on n qubits, which represent elements $\boldsymbol{x} \in \mathbb{F}_2^n$. In the above definition, V_f is the modified query operator working as $V_f |\boldsymbol{x}\rangle = (-1)^{f(\boldsymbol{x})} |\boldsymbol{x}\rangle$, H_n is the Hadamard transform, and R_n is the operator which reverses the sign of $|\mathbf{0}\rangle$, $R_n |\mathbf{0}\rangle = -|\mathbf{0}\rangle$ and $R_n |\boldsymbol{x}\rangle = |\boldsymbol{x}\rangle$ for $\boldsymbol{x} \neq \mathbf{0}$.

It is interesting to see that $H_n R_n H_n$ can be written in quite a simple form as a $2^n \times 2^n$ matrix: if we let elements $\boldsymbol{x} \in \mathbb{F}_2^n$ to index the rows and the columns, we find that, since

$$H_n\left|\boldsymbol{x}\right\rangle = \frac{1}{\sqrt{2^n}}\sum_{\boldsymbol{y}\in\mathbb{F}_2^n}(-1)^{\boldsymbol{x}\cdot\boldsymbol{y}}\left|\boldsymbol{y}\right\rangle,$$

the element of H_n at location $(\boldsymbol{x},\boldsymbol{y})$ is $\frac{1}{\sqrt{2^n}}(-1)^{\boldsymbol{x}\cdot\boldsymbol{y}}$. Writing generally $A_{\boldsymbol{xy}}$ for the element of a $2^n \times 2^n$ matrix A at location $(\boldsymbol{x},\boldsymbol{y})$, we see (recall the matrix representation of R_n in the previous section) that

$$\begin{aligned}
(H_nR_nH_n)_{\boldsymbol{xy}} &= \sum_{\boldsymbol{x}\in\mathbb{F}_2^n}(H_nR_n)_{\boldsymbol{xz}}(H_n)_{\boldsymbol{zy}} \\
&= \sum_{\boldsymbol{z}}\sum_{\boldsymbol{w}}(H_n)_{\boldsymbol{xw}}(R_n)_{\boldsymbol{wz}}(H_n)_{\boldsymbol{zy}} \\
&= \frac{1}{2^n}\sum_{\boldsymbol{z}\in\mathbb{F}_2^n}(-1)^{\boldsymbol{x}\cdot\boldsymbol{z}}(R_n)_{\boldsymbol{zz}}(-1)^{\boldsymbol{z}\cdot\boldsymbol{y}} \\
&= \frac{1}{2^n}(-2+\sum_{\boldsymbol{z}\in\mathbb{F}_2^n}(-1)^{(\boldsymbol{x}+\boldsymbol{y})\cdot\boldsymbol{z}}) \\
&= \begin{cases} -\frac{2}{2^n}, \text{ if } \boldsymbol{x}\neq\boldsymbol{y} \\ 1-\frac{2}{2^n}, \text{ if } \boldsymbol{x}=\boldsymbol{y}. \end{cases}
\end{aligned}$$

Thus $H_nR_nH_n$ has matrix representation

$$H_nR_nH_n = \begin{pmatrix} 1-\frac{2}{2^n} & -\frac{2}{2^n} & \ldots & -\frac{2}{2^n} \\ -\frac{2}{2^n} & 1-\frac{2}{2^n} & \ldots & -\frac{2}{2^n} \\ \vdots & \vdots & \ddots & \vdots \\ -\frac{2}{2^n} & -\frac{2}{2^n} & \ldots & 1-\frac{2}{2^n} \end{pmatrix},$$

which can also be expressed as

$$H_nR_nH_n = I - 2P, \tag{6.18}$$

where I is $2^n \times 2^n$ identity matrix and P is a $2^n \times 2^n$ *projection matrix*, whose every entry is $\frac{1}{2^n}$. In fact, it is quite an easy task to verify that P represents a projection into a one-dimensional subspace generated by vector

$$\psi = \frac{1}{\sqrt{2^n}}\sum_{\boldsymbol{x}\in\mathbb{F}_2^n}\left|\boldsymbol{x}\right\rangle.$$

Thus, using the notations of Chapter 8, we can write $P = \mid\psi\rangle\langle\psi\mid$, but this notation is not crucial here. It is more essential that representation (6.18) gives us an easy method for finding the effect of $-H_nR_nH_n$ on a general superposition

$$\sum_{\boldsymbol{x}\in\mathbb{F}_2^n} c_{\boldsymbol{x}}\left|\boldsymbol{x}\right\rangle. \tag{6.19}$$

Writing

$$A = \frac{1}{2^n} \sum_{x \in \mathbb{F}_2^n} c_x$$

(the average of the amplitudes), we find that

$$\sum_{x \in \mathbb{F}_2^n} c_x \left| x \right\rangle = A \sum_{x \in \mathbb{F}_2^n} \left| x \right\rangle + \sum_{x \in \mathbb{F}_2^n} (c_{x-A}) \left| x \right\rangle \tag{6.20}$$

is the decomposition of (6.19) into two orthogonal vectors: the first belongs to the subspace spanned by ψ; the second belongs to the orthogonal complement of that subspace. In fact, it is clear that the first summand of the right-hand side of (6.20) belongs to the subspace generated by ψ, and the orthogonality of the summands is easy to verify:

$$\begin{aligned} &\langle A \sum_x \left| x \right\rangle \mid \sum_y (c_y - A) \left| y \right\rangle \rangle \\ &= \sum_{x \in \mathbb{F}_2^n} \sum_{y \mathbb{F}_2^n} A^* (c_x - A) \langle x \mid y \rangle \\ &= A^* \sum_{x \in \mathbb{F}_2^n} c_x - \sum_{x \in \mathbb{F}_2^n} A^* A \\ &= A^* 2^n A - 2^n A^* A = 0. \end{aligned}$$

Therefore,

$$P \sum_{x \in \mathbb{F}_2^n} c_x \left| x \right\rangle = A \sum_{x \in \mathbb{F}_2^n} \left| x \right\rangle ,$$

and

$$\begin{aligned} -H_n R_n H_n \sum_{x \in \mathbb{F}_2^n} c_x \left| x \right\rangle &= (2P - I) \sum_{x \in \mathbb{F}_2^n} c_x \left| x \right\rangle \\ &= 2A \sum_{x \in \mathbb{F}_2^n} \left| x \right\rangle - \sum_{x \in \mathbb{F}_2^n} c_x \left| x \right\rangle \\ &= \sum_{x \in \mathbb{F}_2^n} (2A - c_x) \left| x \right\rangle . \end{aligned} \tag{6.21}$$

Expression (6.21) explains why operation $-H_n R_n H_n$ is also called *inversion about average*: it operates on a single amplitude c_x by multiplying it by -1 and adding two times the average.

We use the following example to explain how $G_n = -H_n R_n H_n V_f$ is used to amplify the desired amplitude in order to find the solution (recall that the solution is an element $y \in \mathbb{F}_2^n$ which satisfies $f(y) = 1$). In this example, we consider a function $f_5 : \mathbb{F}_2^n \to \mathbb{F}_2$, which takes value 1 at $y = (0, \ldots, 0, 1, 0, 1)$, and is 0 everywhere else. The search begins with superposition

$$\frac{1}{\sqrt{2^n}} \sum_{x \in \mathbb{F}_2^n} \left| x \right\rangle$$

having uniform amplitudes. Figure 6.1 depicts this superposition, with $c_0 = c_1 = \ldots c_{2^n} = \frac{1}{\sqrt{2^n}}$. The modified query operator V_{f_5} flips the signs of all

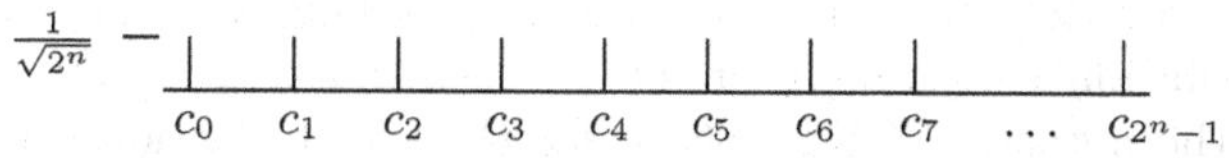

Fig. 6.1. Amplitudes of the initial configuration.

those amplitudes that are coefficients of a vector $|\boldsymbol{x}\rangle$ satisfying $f_5(\boldsymbol{x}) = 1$. In this example, $\boldsymbol{y} = (0, \ldots, 0, 1, 0, 1)$ is the only one having this property, and c_5 becomes $-\frac{1}{\sqrt{2^n}}$. This is depicted in Figure 6.2.

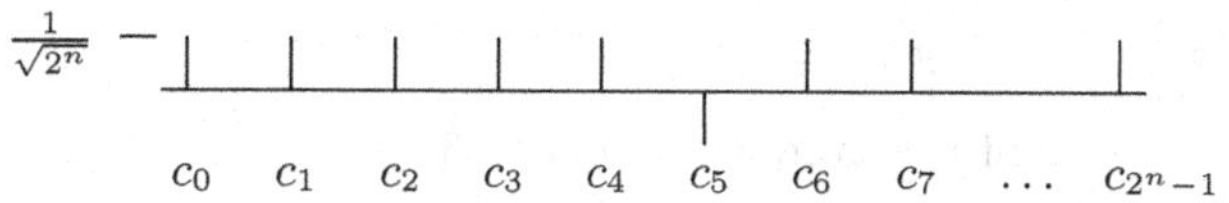

Fig. 6.2. Amplitudes of after one query.

In the case of Figure 6.2, the average of the amplitudes is

$$A = \frac{1}{2^n}\left((2^n - 1)\frac{1}{\sqrt{2^n}} - \frac{1}{\sqrt{2^n}}\right) = \frac{1}{\sqrt{2^n}}\left(1 - \frac{2}{2^n}\right),$$

which is still very close to $\frac{1}{\sqrt{2^n}}$. Thus, the inversion about average-operator $-H_n R_n H_n$ will perform a transformation

$$\begin{cases} \frac{1}{\sqrt{2^n}} \mapsto 2A - \frac{1}{\sqrt{2^n}} \approx \frac{1}{\sqrt{2^n}} \\ -\frac{1}{\sqrt{2^n}} \mapsto 2A + \frac{1}{\sqrt{2^n}} \approx 3 \cdot \frac{1}{\sqrt{2^n}}, \end{cases}$$

which is illustrated in Figure 6.3. It is thus possible to find the required

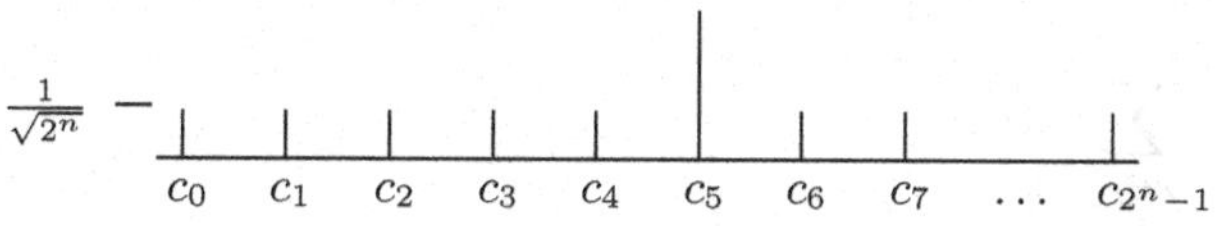

Fig. 6.3. Amplitudes after $-H_n R_n H_n V_f$.

$\boldsymbol{y} = (0, \ldots, 0, 1, 0, 1)$ with a probability $\approx \frac{9}{2^n}$ by just a single query. This is approximately 4.5 times better than a classical randomized search can do.

6.2.3 Analysis of Amplification Method

In this section, we will find out the effect of iterative use of mapping $G_n = -H_n R_n H_n V_f$, but instead of a blackbox function $f : \mathbb{F}_2^n \to \mathbb{F}_2$ that assumes only one solution, we will study a general f having k solutions (recall that here a solution means a vector $\boldsymbol{x} \in \mathbb{F}_2^n$ such that $f(\boldsymbol{x}) = 1$).

Let the notation $T \subseteq \mathbb{F}_2^n$ stand for the set of solutions, and $F = \mathbb{F}_2^n \setminus T$ for the set of non-solutions. Thus $|T| = k$ and $|F| = 2^n - k$. Assume that after r iteration steps, the state is

$$t_r \sum_{\boldsymbol{x}\in T} |\boldsymbol{x}\rangle + f_r \sum_{\boldsymbol{x}\in F} |\boldsymbol{x}\rangle . \tag{6.22}$$

The modified query operator V_f will then give us

$$-t_r \sum_{\boldsymbol{x}\in T} |\boldsymbol{x}\rangle + f_r \sum_{\boldsymbol{x}\in F} |\boldsymbol{x}\rangle , \tag{6.23}$$

and the average of the amplitudes of (6.23) is

$$A = \frac{1}{2^n}\Big(- tk + f(2^n - k)\Big).$$

Operator $-H_n R_n H_n$ will, therefore, transform (6.23) into

$$t_{r+1} \sum_{\boldsymbol{x}\in T} |\boldsymbol{x}\rangle + f_{r+1} \sum_{\boldsymbol{x}\in F} |\boldsymbol{x}\rangle ,$$

where $t_{r+1} = 2A - t_r = (1 - \frac{2k}{2^n})t_r + (2 - \frac{2k}{2^n})f_r$ and $f_{r+1} = 2A - f_r = -\frac{2k}{2^n}t_r + (1 - \frac{2k}{2^n})f_r$. Collecting all of this together, we find that $G_n = -H_n R_n H_n V_f$ operates on (6.22) as a transformation

$$\begin{pmatrix} t_{r+1} \\ f_{r+1} \end{pmatrix} = \begin{pmatrix} 1 - \frac{2k}{2^n} & 2 - \frac{2k}{2^n} \\ -\frac{2k}{2^n} & 1 - \frac{2k}{2^k} \end{pmatrix} \begin{pmatrix} t_r \\ f_r \end{pmatrix} . \tag{6.24}$$

Therefore, to find out the effect of G_n on uniformly weighted initial superposition

$$\frac{1}{\sqrt{2^n}} \sum_{\boldsymbol{x}\in\mathbb{F}_2^n} |\boldsymbol{x}\rangle ,$$

we have to solve (6.24) with boundary conditions $t_0 = f_0 = \frac{1}{\sqrt{2^n}}$. This can be done as follows: it is clear that t_r and f_r are real and, since G_n is a unitary operator, equation

$$kt_r^2 + (2^n - k)f_r^2 = 1 \tag{6.25}$$

must hold for each r. That is to say that each point (t_r, f_r) lies in the ellipse defined by equation (6.25). Therefore, we can write

$$\begin{cases} t_r &= \frac{1}{\sqrt{k}} \sin \theta_r, \\ f_r &= \frac{1}{\sqrt{2^n - k}} \cos \theta_r \end{cases}$$

for some number θ_r. Recursion (6.24) can then be rewritten as

$$\begin{cases} \sin \theta_{r+1} &= \left(1 - \frac{2k}{2^n}\right) \sin \theta_r + \frac{2}{2^n} \sqrt{2^n - k}\sqrt{k} \cos \theta_r, \\ \cos \theta_{r+1} &= -\frac{2}{2^n} \sqrt{2^n - k}\sqrt{k} \sin \theta_r + \left(1 - \frac{2k}{2^n}\right) \cos \theta_r. \end{cases} \tag{6.26}$$

Recall that k is the number of elements in $\mathbb{F}_2^n$ that satisfy $f(\boldsymbol{y}) = 1$, hence, $1 - \frac{2k}{2^n} \in [-1, 1]$. Therefore, we can choose $\omega \in [0, \pi]$ such that $\cos \omega = 1 - \frac{2k}{2^n}$. Then $\sin \omega = \frac{2}{2^n} \sqrt{2^n - k}\sqrt{k}$ and (6.26) becomes

$$\begin{cases} \sin \theta_{r+1} &= \sin(\theta_r + \omega), \\ \cos \theta_{r+1} &= \cos(\theta_r + \omega). \end{cases}$$

The boundary condition gives us that $\sin^2 \theta_0 = \frac{k}{2^n}$, and it is easy to verify that the solution of the recursion is

$$\begin{cases} t_r &= \frac{1}{\sqrt{k}} \sin(r\omega + \theta_0), \\ f_r &= \frac{1}{\sqrt{2^n - k}} \cos(r\omega + \theta_0), \end{cases}$$

where $\theta_0 \in [0, \pi/2]$ and $\omega \in [0, \pi]$ are chosen in such a way that $\sin^2 \theta_0 = \frac{k}{2^n}$ and $\cos \omega = 1 - \frac{2k}{2^n}$. In fact, then $\cos \omega = 1 - 2\sin^2 \phi_0 = \cos 2\theta_0$, so we have obtained the following lemma.

Lemma 6.2.1. *Solution of (6.24) with boundary conditions* $t_0 = f_0 = \frac{1}{\sqrt{2^n}}$ *is*

$$\begin{cases} t_r &= \frac{1}{\sqrt{k}} \sin\left((2r + 1)\theta_0\right), \\ f_r &= \frac{1}{\sqrt{2^n - k}} \cos\left((2r + 1)\theta_0\right), \end{cases} \tag{6.27}$$

where $\theta_0 \in [0, \pi/2]$ *is determined by* $\sin^2 \theta_0 = \frac{k}{2^n}$.

We would like to find a suitable value for r to maximize the probability for finding a solution. Since there are k solutions, the probability of seeing one solution is

$$kt_r^2 = \sin^2 \theta_r = \sin^2((2r + 1)\theta_0). \tag{6.28}$$

We would then like to find a non-negative r as small as possible such that $\sin^2((2r + 1)\theta_0)$ is as close to 1 as possible; i.e., the least positive integer r such that $(2r + 1)\theta_0$ is as close to $\frac{\pi}{2}$ as possible.

We notice that

$$(2r+1)\theta_0 = \frac{\pi}{2} \iff r = -\frac{1}{2} + \frac{\pi}{4\theta_0},$$

and, since $\theta_0^2 \approx \sin^2\theta_0 = \frac{k}{2^n}$, we have

$$\theta_0 \approx \sqrt{\frac{k}{2^n}},$$

and therefore after approximately

$$\left\lfloor \frac{\pi}{4}\sqrt{\frac{2^n}{k}} \right\rfloor$$

iterations, the probability of seeing a desired element $\boldsymbol{y} \in \mathbb{F}_2^n$ is quite close to 1. To be more precise:

Theorem 6.2.1. *Let $f : \mathbb{F}_2^n \to \mathbb{F}_2$ such that there are k elements $\boldsymbol{x} \in \mathbb{F}_2^n$ satisfying $f(\boldsymbol{x}) = 1$. Assume that $0 < k \leq \frac{3}{4} \cdot 2^n$,*[1] *and let $\theta_0 \in [0, \pi/3)$ be chosen such that $\sin^2\theta_0 = \frac{k}{2^n} \leq \frac{3}{4}$. After $\left\lfloor \frac{\pi}{4\theta_0} \right\rfloor$ iterations of G_n on an initial superposition*

$$\frac{1}{\sqrt{2^n}} \sum_{\boldsymbol{x}\in\mathbb{F}_2^n} |\boldsymbol{x}\rangle$$

the probability of seeing a solution is at least $\frac{1}{4}$.

Proof. The probability of seeing a desired element is given by $\sin^2((2r+1)\theta_0)$, and we just saw that $r = -\frac{1}{2} + \frac{\pi}{4\theta_0}$ would a give of probability 1. Thus, we only have to estimate the error when $-\frac{1}{2} + \frac{\pi}{4\theta_0}$ is replaced by $\left\lfloor \frac{\pi}{4\theta_0} \right\rfloor$.

Clearly

$$\left\lfloor \frac{\pi}{4\theta_0} \right\rfloor = -\frac{1}{2} + \frac{\pi}{4\theta_0} + \delta$$

for some δ with $|\delta| \leq \frac{1}{2}$. Therefore,

$$\left(2\left\lfloor \frac{\pi}{4\theta_0} \right\rfloor + 1\right)\theta_0 = \frac{\pi}{2} + 2\delta\theta_0,$$

so the distance of $(2\left\lfloor \frac{\pi}{4\theta_0} \right\rfloor + 1)\theta_0$ from $\frac{\pi}{2}$ is $|2\delta\theta_0| \leq \frac{\pi}{3}$. It follows that

$$\sin^2\left(2\left\lfloor \frac{\pi}{4\theta_0} \right\rfloor + 1\right)\theta_0 \geq \sin^2\left(\frac{\pi}{2} - \frac{\pi}{3}\right) = \frac{1}{4}.$$

[1] If $k > \frac{3}{4} \cdot 2^n$, we can find a desired solution $\boldsymbol{x}$ with a probability of at least $\frac{3}{4}$ just by guessing, and if $k = 0$, then G_n does not alter the initial superposition at all.

Grover's method for quantum search is summarized in the following algorithm:

Grover's Search Algorithm

Input: A blackbox function $f : \mathbb{F}_2^n \to \mathbb{F}_2$ and $k = |\{\boldsymbol{x} \in \mathbb{F}_2^n \mid f(\boldsymbol{x}) = 1\}|$.

Output: An $\boldsymbol{y} \in \mathbb{F}_2^n$ such that $f(\boldsymbol{y}) = 1$ (a solution), if such an element exists.

1. If $k > \frac{3}{4} \cdot 2^n$, choose $\boldsymbol{y} \in \mathbb{F}_2^n$ with uniform probability and stop.
2. Otherwise compute $r = \left\lfloor \frac{\pi}{4\theta_0} \right\rfloor$, where $\theta_0 \in [0, \pi/3]$ is determined by $\sin^2 \theta_0 = \frac{k}{2^n}$.
3. Prepare the initial superposition
$$\frac{1}{\sqrt{2^n}} \sum_{\boldsymbol{x} \in \mathbb{F}_2^n} |\boldsymbol{x}\rangle$$
by using Hadamard transform H_n.
4. Apply operator G_n r times.
5. Observe to get some $\boldsymbol{y} \in \mathbb{F}_2^n$.

If $k \geq \frac{3}{4} \cdot 2^n$, then the first step clearly produces correct output with a probability of at least $\frac{3}{4}$. Otherwise, according to Theorem (6.2.1), the algorithm gives a solution with a probability of at least $\frac{1}{4}$. In each case, we can find a solution with nonvanishing probability.

If $k = 1$ and n large, then $\left\lfloor \frac{\pi}{4\theta_0} \right\rfloor \approx \frac{\pi}{4}\sqrt{2^n}$, so using $O(\sqrt{2^n})$ queries to f we can find a solution with nonvanishing probability, which is essentially better than any classical randomized algorithm can do.

Case $k = \frac{1}{4} \cdot 2^n$ is also very interesting. Then, $\sin^2 \theta_0 = \frac{1}{4}$, so $\theta_0 = \frac{\pi}{6}$. According to (6.28), the probability of seeing a solution after one single iteration of G_n is

$$\sin^2(3\theta_0) = \sin^2 \frac{\pi}{2} = 1.$$

Thus for $k = \frac{1}{4} \cdot 2^n$ we can find a solution *with certainty* using G_n only once, which is clearly impossible using any classical search strategy.

In a typical situation we, unfortunately, do not know the value of k in advance. In the next section, we present a simplified version of a method due to M. Boyer, G. Brassard, P. Høyer and A. Tapp [17] in order to find the required element even if k is not known.

6.3 Utilizing Grover's Search Method

6.3.1 Searching with Unknown Number of Solutions

We begin with an elementary lemma, whose proof is left as an exercise.

Lemma 6.3.1. *For any real α and any positive integer m,*

$$\sum_{r=0}^{m-1} \cos((2r+1)\alpha) = \frac{\sin(2m\alpha)}{2\sin\alpha}.$$

The following important lemma can be found in [17].

Lemma 6.3.2. *Let $f : \mathbb{F}_2^n \to \mathbb{F}_2$ a blackbox function with $k \le \frac{3}{4} \cdot 2^n$ solutions and $\theta_0 \in [0, \frac{\pi}{3}]$ defined by equation $\sin^2\theta_0 = \frac{k}{2^n}$. Let m be any positive integer and $r \in [0, m-1]$ chosen with uniform distribution. If G_n is applied to initial superposition*

$$\frac{1}{\sqrt{2^n}} \sum_{\boldsymbol{x} \in \mathbb{F}_2^n} |\boldsymbol{x}\rangle$$

r times, then the probability of seeing a solution is

$$P_r = \frac{1}{2} - \frac{\sin(4m\theta_0)}{4m\sin(2\theta_0)}.$$

Proof. In the previous section we saw that, after r iterations of G_n, the probability of seeing a solution is $\sin^2((2r+1)\theta_0)$. Thus, if $r \in [0, m-1]$ is chosen uniformly, then the probability of seeing a solution is

$$\begin{aligned} P_m &= \frac{1}{m} \sum_{r=0}^{m-1} \sin^2((2r+1)\theta_0) \\ &= \frac{1}{2m} \sum_{r=0}^{m-1} (1 - \cos((2r+1)2\theta_0)) \\ &= \frac{1}{2} - \frac{\sin(4m\theta_0)}{4m\sin(2\theta_0)} \end{aligned}$$

according to Lemma 6.3.1. □

Remark 6.3.1. If $m \ge \frac{1}{\sin(2\theta_0)}$, then

$$\sin(4m\theta_0) \le 1 = \frac{1}{\sin(2\theta_0)} \sin(2\theta_0) \le m \sin(2\theta_0),$$

and therefore $\frac{\sin(4m\theta_0)}{4m\sin(2\theta_0)} \le \frac{1}{4}$. The above lemma implies, then, that $P_r \ge \frac{1}{4}$. This means that if m is great enough, then applying G_n r times, where $r \in [0, m-1]$ is chosen uniformly, will yield a solution anyway with a probability of at least $\frac{1}{4}$.

Assume now that the unknown number k satisfies $0 < k \le \frac{3}{4} \cdot 2^n$. Then

$$\frac{1}{\sin(2\theta_0)} = \frac{1}{2\sin\theta_0\cos\theta_0} = \frac{2^n}{2\sqrt{k(2^n-k)}} \le \sqrt{\frac{2^n}{k}} \le \sqrt{2^n},$$

so choosing $m \geq \sqrt{2^n}$ we certainly have $m \geq \frac{1}{\sin(2\theta_0)}$. This leads to the following algorithm:

Quantum Search Algorithm

Input: A blackbox function $f : \mathbb{F}_2^n \to \mathbb{F}_2$.

Output: Any $\boldsymbol{y} \in \mathbb{F}_2^n$ such that $f(\boldsymbol{y}) = 1$ (a solution), if such an element exists.

1. Pick an element $\boldsymbol{x} \in \mathbb{F}_2^n$ randomly with uniform distribution. If $f(\boldsymbol{x}) = 1$, then output $\boldsymbol{x}$ and stop.
2. Otherwise, let $m = \lfloor \sqrt{2^n} \rfloor + 1$ and choose an integer $r \in [0, m-1]$ uniformly.
3. Prepare the initial superposition

$$\frac{1}{\sqrt{2^n}} \sum_{\boldsymbol{x} \in \mathbb{F}_2^n} |\boldsymbol{x}\rangle$$

by using Hadamard transform H_n and apply G_n r times.
4. Observe to get some $\boldsymbol{y} \in \mathbb{F}_2^n$.

Using the previous lemmata, the correctness probability of this algorithm is easy to analyze: let k be the number of solutions to f. If $k > \frac{3}{4} \cdot 2^n$, then the algorithm will output a solution after the first instruction with a probability of at least $\frac{3}{4}$. On the other hand, if $k \leq \frac{3}{4} \cdot 2^n$, then

$$m \geq \sqrt{\frac{2^n}{k}} \geq \frac{1}{\sin(2\theta_0)},$$

and, according to Remark 6.3.1, the probability of seeing a solution is at least $\frac{1}{4}$ anyway.

Remark 6.3.2. Since the above algorithm is guaranteed to work with a probability of at least $\frac{1}{4}$, we can say that on average the solution can be found after four attempts. In each attempt, the number of queries to f is at most $\sqrt{2^n}$.

A linear improvement of the above algorithm can be obtained by emplying the methods of [17]. In any case, the above algorithm results in the following theorem.

Theorem 6.3.1. *By using a quantum circuit making $O(\sqrt{2^n})$ queries to a blackbox function f, one can decide with nonvanishing correctness probability if there is an element $\boldsymbol{x} \in \mathbb{F}_2^n$ such that $f(\boldsymbol{x}) = 1$.*

In the next chapter we will present a clever technique formulated by R. Beals, H. Buhrman, R. Cleve, M. Mosca, and R. de Wolf [6], which will allow us to demonstrate that the above algorithm is, in fact, optimal up to a constant factor. Any quantum algorithm that can discover, with nonvanishing probability, whether a solution to f exists uses $\Omega(\sqrt{2^n})$ queries to f. This result concerns blackbox functions, so it does not imply a complexity lower bound for computable functions (see Remark 6.1.3).

On the other hand, blackbox function f can be replaced with a computable function to get the following theorem.

Theorem 6.3.2. *By using a quantum circuit, any problem in* **NP** *can be solved with nonvanishing correctness probability in time* $O(\sqrt{2^n}p(n))$, *where* p *is a polynomial depending on the particular problem.*

In the above theorem, polynomial $p(n)$ is essentially that the nondeterministic computation needs to solve the problem (see Section 3.1.2).

Grover's search algorithm has been cunningly used for several purposes. To mention a few examples, see [19] for quantum counting algorithm, and [35] for minimum finding algorithm. In [17] the authors outline a generalized search algorithm for sets other than $\mathbb{F}_2^n$.

Remark 6.3.3. Recently, L. Grover and T. Rudolph have argued [42] that many algoritms based on the "standard quantum search method" fail to be nontrivial in the sense that the speedup provided could be obtained just as well by dividing the search space into suitable parts, and then performing the "standard search" in parallel.

7. Complexity Lower Bounds for Quantum Circuits

Unfortunately, the most interesting questions related to quantum complexity theory, for example, "Is **NP** contained in **BQP** or not?", are extremely difficult and far beyond recent knowledge. In this chapter we will mention some complexity theoretical aspects of quantum computation, which may somehow illustrate these difficult questions.

7.1 General Idea

A general technique for deriving complexity theoretical lower bounds for quantum circuits using blackbox function queries was introduced by R. Beals, H. Buhrman, R. Cleve, M. Mosca, and R. de Wolf in [6]. Before going into the details, let us briefly discuss the idea informally. To simplify some notations, we will sometimes interpret an element $\boldsymbol{x} \in \mathbb{F}_2^n$ as a binary number in interval $[0, 2^n - 1]$.

Assume that a quantum circuit finds out if an arbitrary blackbox function

$$f : \mathbb{F}_2^n \to \mathbb{F}_2$$

has some property or not (for instance, in Section 6.1.2 we were interested in whether there was an element $\boldsymbol{y} \in \mathbb{F}_2^n$ such that $f(\boldsymbol{y}) = 1$). This quantum circuit can then be seen as a device which is given vector $(f(0), f(1), \ldots, f(2^n - 1))$ (the values of f) as an input, and it gives output "yes" or "no" respectively if f has the property or not. In other words, the quantum circuit can then be seen as a device for computing a $\{0, 1\}$-valued function on input vector $(f(0), f(1), \ldots, f(2^n - 1))$ (unknown to us), which is to say that the *circuit is a device for computing a Boolean function on* 2^n *input variables* $f(0)$, $f(1)$, …, $f(2^n - 1)$.

Although lot of things about Boolean functions are unknown to us, fortunately, we do know many things about them; see [64] for discussion. It turns out that the *polynomial representations* of Boolean functions will give us the desired information about the number of blackbox queries needed to find out how many queries are required to find out some property of f. The bound will be expressed in terms of *polynomial representation degree* for quantum circuits that exactly (with a probability of 1) compute the property (Boolean function), and in terms of *polynomial approximation degree* for circuits that have a high probability of finding that property. We will see that the number of queries to the blackbox function will bound above the representation degree of a Boolean function computable with a particular quantum circuit. The lower bound for the number of necessary queries is thus established by finding the representation degree of the Boolean function to be computed: the quantum circuit must make enough queries to reach the representation degree. Because the polynomial representations and approximations play such a key role, we will devote the following two sections to study them.

7.2 Polynomial Representations

7.2.1 Preliminaries

In this section, we represent the basic facts about polynomial representations of Boolean functions and symmetric polynomials. A reader who is already familiar with these topics may skip this section.

Let $N = 2^n$. As discussed in Sections 4.1.1 and 9.2.2, functions

$$F : \mathbb{F}_2^N \to \mathbb{C}$$

form an 2^N-dimensional vector space V over complex numbers. The idea is that Boolean functions on N variables can be viewed as elements of V: they are functions defined in N binary variables getting values in a two-element set $\{0,1\}$ which can be embedded in $\mathbb{C}$.

For each subset $K = \{k_1, k_2, \ldots, k_l\}, \subseteq \{0, 1, \ldots, N-1\}$ we consider a function

$$X_K : \mathbb{F}_2^N \to \mathbb{C}, \tag{7.1}$$

which we denote also by $X_K = X_{k_1} X_{k_2} \ldots X_{k_l}$ and define in a most natural way: the value of function X_K on a vector $\boldsymbol{x} = (x_0, x_1, \ldots, x_{N-1})$ is the product $x_{k_1} x_{k_2} \cdot \ldots \cdot x_{k_l}$, which is interpreted as the number 0 or 1. This may require some explanation: $\boldsymbol{x}$ is an element in $\mathbb{F}_2^N$, and each of its component x_i is an element of the binary field $\mathbb{F}_2$. Thus, the product $x_{k_1} x_{k_2} \cdot \ldots \cdot x_{k_l}$ also belongs to $\mathbb{F}_2$, but embedding $\mathbb{F}_2$ in $\mathbb{C}$, we may interpret the product as also belonging to $\mathbb{C}$. It is natural to call functions X_K *monomials* and to denote $X_\emptyset = 1$ (an empty product having no factors at all is interpreted as 1). If $|K| = l$, the *degree* of a monomial $X_K = X_{k_1} X_{k_2} \ldots X_{k_l}$ is defined to be $\deg X_K = l$. A linear combination of monomials X_{k_1}, ..., X_{K_s} is naturally denoted by

$$P = c_1 X_{k_1} + \ldots + c_s X_{K_s} \tag{7.2}$$

and is called a polynomial. The degree $\deg P$ of polynomial (7.2) is defined as the highest degree of a monomial occurring in (7.2), except in the case of zero polynomial $p = 0$, whose degree we symbolically define to be $-\infty$.

A reader may now wonder why we would consider such a simple concept as a polynomial so precisely? The problematic point is that, in a finite field, the same function can be defined by many different polynomials. For example, both elements of the binary field $\mathbb{F}_2$ satisfy equation $x^2 = x$ and, thus, non-constant polynomial $x^2 - x + 1$ behaves like a constant polynomial 1.

We will, therefore, define the product of monomials X_K and X_L by considering X_K and X_L as functions $\mathbb{F}_2^N \to \mathbb{F}_2$. It is plain to see that this will result in definition $X_K X_L = X_{K \cup L}$. The product of two polynomials is defined as usual: if

$$P_1 = c_1 X_{k_1} + c_2 X_{k_2} + \ldots + c_s X_{K_s}$$

and

$$P_2 = d_1 X_{l_1} + d_2 X_{l_2} + \ldots + c_t X_{l_t}$$

are polynomials, we define

$$P_1 P_2 = \sum_{i=1}^{s} \sum_{j=1}^{t} c_i d_j X_{k_i} X_{l_j}.$$

The product differs slightly from the ordinary product of polynomials. Take, for example, $K = \{0, 1\}$ and $L = \{1, 2\}$. Then

$$X_K X_L = X_0 X_1 \, X_1 X_2 = X_0 X_1 X_2 = X_{\{0,1,2\}},$$

but if $X_0 X_1$ and $X_1 X_2$ were seen as "ordinary polynomials" over F_2, the product would have been $X_0 X_1^2 X_2$. In fact, the The functions behave just like ordinary polynomials, except all powers above 1 must be reduced to 1.[1]

Lemma 7.2.1. *The monomials form a basis of* V.

Proof. For any $\boldsymbol{y} = (y_0, y_1, \ldots, y_{2^n-1}) \in \mathbb{F}_2^N$, define a polynomial

$$P_{\boldsymbol{y}} = (1 + y_0 - X_0)(1 + y_1 - X_1) \ldots (1 + y_{N-1} - X_{N-1}).$$

By expanding the product, we can get a representation for $P_{\boldsymbol{y}}$ as a sum of monomials, and for any $\boldsymbol{x} = (x_0, x_1, \ldots, x_{N-1}) \in \mathbb{F}_2^N$,

$$\begin{aligned} & P_{\boldsymbol{y}}(\boldsymbol{x}) = 1 \\ \iff & 1 + y_i - x_i = 1 \text{ for each } i \\ \iff & x_i = y_i \text{ for each } i \\ \iff & \boldsymbol{x} = \boldsymbol{y}, \end{aligned}$$

[1] For a reader who is aware of algebraic structures: the polynomial functions considered here form a ring isomorphic to $\mathbb{F}_2[X_0, X_1, \ldots, X_{N-1}]/I$, where I is the ideal generated by all polynomials $X_i^2 - X_i$.

which is to say that $P_{\boldsymbol{y}}$ is the *characteristic function* of singleton set $\{\boldsymbol{y}\}$. But because we can express each characteristic function $\mathbb{F}_2^N \to \mathbb{C}$ as a sum of monomials, we can surely express *any* function $\mathbb{F}_2^N \to \mathbb{C}$ as a linear combination of monomials (see Section 9.2.2). Therefore, monomials generate V, and since there are

$$\binom{N}{0} + \binom{N}{1} + \ldots + \binom{N}{N} = 2^N$$

different monomials, they are also linearly independent. □

It follows from the above lemma that each function $\mathbb{F}_2^N \to \mathbb{F}_2$ can be uniquely expressed as a linear combination of monomials (7.1) with complex coefficients. Clearly, all Boolean functions with N variables are included. In fact, we would have also been able to derive such a representation for Boolean functions also directly: Boolean variables can be represented by monomials X_0, ..., X_{N-1}, and if P_1 and P_2 are polynomial representations of some Boolean functions, then $\neg P_1$ can be represented as $1 - P_1$, $P_1 \wedge P_2$ as $P_1 P_2$, and $P_1 \vee P_2$ as $1 - (1 - P_1)(1 - P_2)$. Since all Boolean functions can be constructed by using $\neg$, $\wedge$, and $\vee$, (see Section 3.2.1) we have the following lemma.

Lemma 7.2.2. *Boolean functions on N variables have a unique representation as a multivariate polynomial $P(X_0, X_1, \ldots X_{N-1})$ having real coefficients and degree at most N.*

We must keep in mind that "polynomial" in the above lemma is interpreted in such a way that $X^k = X$ for each $k \geq 1$.

In the continuation we will mainly concentrate on a subclass of *symmetric* Boolean functions that form a subclass of symmetric functions $F : \mathbb{F}_2^N \to \mathbb{C}$. Symmetric functions are formally defined as follows:

Definition 7.2.1. *Let $\boldsymbol{x} = (x_0, x_1, \ldots, x_{N-1}) \in \mathbb{F}_2^N$ be any vector and $\pi : \{0, 1, \ldots, N-1\} \to \{0, 1, \ldots, N-1\}$ any permutation. Denote $\pi(\boldsymbol{x}) = (x_{\pi(0)}, x_{\pi(1)}, \ldots, x_{\pi(N-1)})$. A function $F : \mathbb{F}_2^N \to \mathbb{C}$ is symmetric if for any vector $\boldsymbol{x} \in \mathbb{F}_2^N$ and for any permutation $\pi \in S_N$,*

$$F(\boldsymbol{x}) = F(\pi(\boldsymbol{x}))$$

holds.

Clearly, any linear combination of symmetric functions is again symmetric, thus, symmetric functions form a subspace of $W \subset V$. We will now find a basis for W. Let P be a nonzero polynomial representing a symmetric function. Then P can be represented as a sum of *homogenous* polynomials[2]

[2] A homogenous polynomial is a linear combination of monomials having the same degree.

$$P = Q_0 + Q_1 + Q_2 + \ldots + Q_N$$

such that $\deg Q_i = i$. Because P itself was assumed to be symmetric, each Q_i must also be a symmetric polynomial (otherwise, by permuting the variables, we could get a different polynomial representing the same function as P does, which is impossible). Thus, we can write

$$Q_i = \sum c_{k_1,k_2,\ldots,k_i} X_{k_1} X_{k_2} \ldots X_{k_i},$$

where the sum is taken over all subsets of $\{0, 1, \ldots, N-1\}$ having a cardinality of i. Because Q_i is invariant under variable permutations, and because representation as a polynomial is unique, we must have that $c_{k_1,k_2,\ldots,k_i} = c_i$ independent of the choice of $\{k_1, k_2, \ldots, k_i\}$. That is to say, symmetric polynomials

$$\begin{aligned}
V_0 &= 1, \\
V_1 &= X_0 + X_1 + \ldots + X_{N-1}, \\
V_2 &= X_0X_1 + X_0X_2 + \ldots + X_{n-2}X_{N-1} \\
\ldots & \quad \ldots \\
V_N &= X_0X_1 \ldots X_{N-1}
\end{aligned}$$

generate W. Because they have different degrees, they are also linearly independent.

The following triviality is worth mentioning : the value of a symmetric function $F(\boldsymbol{x})$ depends only on the *Hamming weight* of $\boldsymbol{x}$, which is given by

$$\mathrm{wt}(\boldsymbol{x}) = |\{i \mid x_i = 1\}| \,.$$

We can even quite easily find out the value of $V_i(\boldsymbol{x})$ when $\mathrm{wt}(\boldsymbol{x}) = k$: consider a polynomial

$$P(T) = (T - X_0)(T - X_1) \ldots (T - X_{N-1}).$$

Since P is invariant under any permutation of X_0, X_1, ..., X_{N-1}, the coefficients of $P(T)$ are symmetric polynomials in X_0, X_1, ..., X_{N-1}. In fact, by expanding P, we find out that the coefficient of T^{N-i} is exactly $(-1)^i V_i$. On the other hand, by substituting 1 for k variables X_j and 0 for the rest, $P(T)$ becomes

$$P(T) = T^{N-k}(T-1)^k = \sum_{i=0}^{k} \binom{k}{i} T^{N-l}(-1)^i,$$

which tells us that $V_i(\boldsymbol{x}) = \binom{k}{i}$ if $i \leq k$, and $V_i(\boldsymbol{x}) = 0$ if $i > k$. This leads us to an important observation which will be used later.

Lemma 7.2.3. *If P is a symmetric polynomial of N variables having complex (resp. real) coefficients and degree d, then there is a polynomial P_w on a single variable with complex (resp. real) coefficients such that*

$$P_w(\mathrm{wt}(\boldsymbol{x})) = P(\boldsymbol{x})$$

for each $\boldsymbol{x} \in \mathbb{F}_2^N$.

Proof. Because P is symmetric, it can be represented as

$$P = c_0 V_0 + c_1 V_1 + \ldots c_d V_d.$$

The claim follows from the fact that $V_i(\boldsymbol{x}) = \binom{\mathrm{wt}(\boldsymbol{x})}{d}$ is a polynomial of $\mathrm{wt}(\boldsymbol{x})$ having degree i. □

7.2.2 Bounds for the Representation Degrees

There are many complexity measures for a Boolean function: the number of Boolean gates $\neg$, $\wedge$, and $\vee$ needed to implement the function (Section 3.2.1) and decision tree complexity, to mention a few. For our purposes, the ideal complexity measure is the degree of the multivariate polynomial representing the function. We follow the notations of [6] in the definition below.

Definition 7.2.2. *Let $B : \mathbb{F}_2^N \to \{0,1\}$ be a Boolean function on N variables. The polynomial representation degree* $\deg(B)$ *of B is the degree of the unique multivariate polynomial representing B.*

For future use, we will define *polynomial approximations* of a Boolean function. Since the interesting results about approximating pertain to polynomials with real coefficients, we will also restrict ourselves to polynomials with real coefficients.

Definition 7.2.3. *A multivariate polynomial P having real coefficients approximates a Boolean function $B : \mathbb{F}_2^N \to \{0,1\}$ if*

$$|B(\boldsymbol{x}) - P(\boldsymbol{x})| \leq \frac{1}{3}$$

for each $\boldsymbol{x} \in \mathbb{F}_2^N$.

Clearly, there may be many polynomials which approximate a single Boolean function B, but we are interested in the minimum degree of an approximating polynomial. Therefore, following [6], we give the following definition.

Definition 7.2.4. *The polynomial approximation degree of a Boolean function B is*

$$\widetilde{\deg}(B) = \min\{\deg(P) \mid P \textit{ is a polynomial approximating } B\}.$$

It is a nontrivial task to find bounds for the polynomial representation degree, but we can derive a general bound for symmetric Boolean functions (for more bounds, see and [6] and the references within). The following lemma can be found in [6]:

Lemma 7.2.4. *Let* $B : \mathbb{F}_2^N \to \{0,1\}$ *be a non-constant symmetric Boolean function. Then* $\deg(B) \geq N/2 + 1$.

Proof. Let P be the symmetric multivariate polynomial of $\deg(B)$ that represents B. By Lemma 7.2.3, there is a single-variate polynomial P_w such that $P_w(\mathrm{wt}(\boldsymbol{x})) = P(\boldsymbol{x})$ for each $\boldsymbol{x} \in \mathbb{F}_2^N$. As a representation of a Boolen function B, $P(\boldsymbol{x})$ is either 0 or 1, and since there are $N+1$ possible Hamming weights (from 0 to N), and P is not constant, $P_w(\mathrm{wt}(\boldsymbol{x})) = 0$ for at least $N/2+1$ integers or $P_w(\mathrm{wt}(\boldsymbol{x})) = 1$ for at least $N/2+1$ integers $\mathrm{wt}(\boldsymbol{x})$. In the former case, we conclude that $\deg(P_w) \geq N/2+1$, and in the latter, we conclude that $\deg(P_w) = \deg(P_w - 1) \geq N/2+1$. Altogether, we find that

$$\deg(B) = \deg(P) = \deg(P_w) \geq N/2 + 1.$$

□

For some particular functions, we can, of course, get better results than the previous theorem gives. Consider, for instance, function $\mathrm{AND} : \mathbb{F}_2^N \to \{0,1\}$ defined as $\mathrm{AND}(\boldsymbol{x}) = 1$ if and only if $\mathrm{wt}(\boldsymbol{x}) = N$. Clearly AND is a symmetric function, and the corresponding single-variate polynomial AND_w representing AND has N zeros $\{0, 1, \ldots, N-1\}$. Therefore, $\deg(\mathrm{AND}) = N$. Using a similar argument, we also find that $\deg(\mathrm{OR}) = N$, where OR is a function taking a value of 0 if and only if $\mathrm{wt}(\boldsymbol{x}) = 0$.

It appears much more difficult to find bounds for the degree of an approximating polynomial. R. Paturi has proved [67] the following theorem which characterizes the degree of a polynomial approximating symmetric Boolean functions. To state Paturi's theorem, we fix some notations. Let $B : \mathbb{F}_2^N \to \{0,1\}$ be a symmetric Boolean function. Then B can be represented as a single-variate polynomial $B_w(\mathrm{wt}(\boldsymbol{x})) = B(\boldsymbol{x})$ depending only on the Hamming weight of $\boldsymbol{x}$. Let

$$\Gamma(B) = \min\{|2k - N + 1| \mid B_w(k) \neq B_w(k+1) \text{ and } 0 \leq k \leq N-1\}.$$

Thus, $\Gamma(B)$ measures how close to weight $N/2$ function B_w changes value: if $B_w(k) \neq B_w(k+1)$ for some k approximately $N/2$, then $\Gamma(B)$ is low. Recall that $B_w(k) \neq B_w(k+1)$ means that, if $\mathrm{wt}(\boldsymbol{x}) = k$, then value $B(\boldsymbol{x})$ changes if we flip one more coordinate of $\boldsymbol{x}$ to 1.

Example 7.2.1. Consider function OR, whose only change occurs when the weight of the argument increases from 0 to 1. Thus, $\Gamma(\mathrm{OR}) = N-1$. Similarly we see that the only jump for function AND occurs when the weight of $\boldsymbol{x}$ increases from $N-1$ to N. Therefore, we have that $\Gamma(\mathrm{AND}) = N-1$.

Example 7.2.2. Let PARITY : $\mathbb{F}_2^N \to \{0,1\}$ be a function defined as PARITY($\boldsymbol{x}$) = 1, if wt($\boldsymbol{x}$) is divisible by 2 and 0 otherwise. Function PARITY always changes its value when the Hamming weight of $\boldsymbol{x}$ increases. For our choice $N = 2^n$ is even, so $\Gamma(PARITY) = 1$, but the theorem also holds if N is odd, and then $\Gamma(PARITY) = 0$.

Function MAJORITY : $\mathbb{F}_2^N \to \{0,1\}$ is defined to take a value of 0, if $\mathrm{wt}(\boldsymbol{x}) < N/2$, and 1 if $\mathrm{wt}(\boldsymbol{x}) \geq N/2$. For even N, the only jump occurs when the weight of $\boldsymbol{x}$ passes from $N/2 - 1$ to $N/2$, and for odd N when wt($\boldsymbol{x}$) increases from $(N-1)/2$ to $(N+1)/2$; so, $\Gamma(\text{MAJORITY}) = 1$ for even N and 0 for odd N.

Theorem 7.2.1 (R. Paturi, [67]). *Let $B : \mathbb{F}_2^n \to \{0,1\}$ be a non-constant symmetric Boolean function. Then*

$$\widetilde{\deg}(B) = \Theta(\sqrt{N(N - \Gamma(B))}).$$

Earlier, we saw that deg(OR) = deg(AND) = N, which means that an exact representation of OR and AND requires a polynomial of degree N. On the other hand, by using previous examples and Paturi's theorem, we find that a polynomial *approximating* these functions may have essentially lower degree: $\widetilde{\deg}(\text{OR}) = \Theta(\sqrt{N})$, and $\widetilde{\deg}(\text{AND}) = \Theta(\sqrt{N})$. However, $\widetilde{\deg}(\text{PARITY}) = \Theta(N)$ and $\widetilde{\deg}(\text{MAJORITY}) = \Theta(N)$, so even an approximating polynomial for these two latter functions has degree $\Omega(N)$.

7.3 Quantum Circuit Lower Bound

7.3.1 General Lower Bound

We are now ready to present the idea of [6], which connects quantum circuits computing Boolean functions and polynomials representing and approximating Boolean functions.

We consider blackbox functions $f : \mathbb{F}_2^n \to \mathbb{F}_2$, and as earlier, quantum blackbox queries are modelled by using a query operator Q_f operating as

$$Q_f \left|\boldsymbol{x}\right\rangle \left|b\right\rangle = \left|\boldsymbol{x}\right\rangle \left|b \oplus f(\boldsymbol{x})\right\rangle.$$

There may also be more qubits for other computational purposes, but without violating generality, we may assume that a blackbox query takes place on fixed qubits. Therefore, we can assume that a general state of a quantum circuit computing a property of f is a superposition of states of form

$$\left|\boldsymbol{x}\right\rangle \left|b\right\rangle \left|\boldsymbol{w}\right\rangle, \tag{7.3}$$

where the first register has n qubits, which are used as source bits of a blackbox query, where $b \in \mathbb{F}_2$ is the source qubit, and where $\boldsymbol{w}$ is a string of some number, say m qubits, which are needed for other computational purposes. There are 2^{n+m+1} different states (7.3), and we can expand the definition of Q_f to include all states (7.3) by operating on $|\boldsymbol{w}\rangle$ as an identity operator; but, we will just denote the enlargened operator again by Q_f:

$$Q_f \, |\boldsymbol{x}\rangle \, |b\rangle \, |\boldsymbol{w}\rangle = |\boldsymbol{x}\rangle \, |b \oplus f(\boldsymbol{x})\rangle \, |\boldsymbol{w}\rangle \, .$$

We begin by finding the eigenvectors of Q_f. It is easy to verify that, if s is either 0 or 1, then

$$\begin{aligned} &Q_f\big(|\boldsymbol{x}\rangle \, |0\rangle \, |\boldsymbol{w}\rangle + (-1)^s \, |\boldsymbol{x}\rangle \, |1\rangle \, |\boldsymbol{w}\rangle \big) \\ &= (-1)^{f(\boldsymbol{x})\cdot s}\big(|\boldsymbol{x}\rangle \, |0\rangle \, |\boldsymbol{w}\rangle + (-1)^s \, |\boldsymbol{x}\rangle \, |1\rangle \, |\boldsymbol{w}\rangle \big), \end{aligned} \tag{7.4}$$

which means that

$$|\boldsymbol{x}\rangle \, |0\rangle \, |\boldsymbol{w}\rangle + (-1)^s \, |\boldsymbol{x}\rangle \, |1\rangle \, |\boldsymbol{w}\rangle \tag{7.5}$$

is an eigenvector of Q_f belonging to an eigenvalue $(-1)^{f(\boldsymbol{x})\cdot s}$. On the other hand, vector (7.5) can be chosen in 2^{n+m+1} different ways and it is simple to verify that they form an orthogonal set, so (7.5) are all the eigenvectors of Q_f. It is clear that if the quantum circuit uses query operator Q_f, the state of the quantum circuit depends on the values of f, but equation (7.4) tells us the dependence in a more explicit manner: Q_f introduces a factor $(-1)^{f(\boldsymbol{x})\cdot s}$, which has a polynomial representation $1 - 2sf(\boldsymbol{x})$. The next lemma from [6] expresses this dependency more precisely.

Lemma 7.3.1. *Let $\mathcal{Q}$ be a quantum circuit that makes T queries to a blackbox function $f : \mathbb{F}_2^n \to \mathbb{F}_2$. Let $N = 2^n$ and identify each binary string in $\mathbb{F}_2^n$ with the number it represents. Then the final state of the circuit is a superposition of states (7.3), whose coefficients are functions of values $X_0 = f(0)$, $X_1 = f(1)$, ..., $X_{N-1} = f(N-1)$. These functions have a polynomial representation degree of at most T.*

Proof. The computation of $\mathcal{Q}$ can be viewed as a sequence of unitary transformations

$$U_0, Q_f, U_1, Q_f, U_2, \ldots, U_{T-1}, Q_f, U_T$$

in a complex vector space spanned by all states (7.3). Operators U_i are fixed but Q_f, of course, depends on the particular blackbox function f, so the coefficients are also functions of X_0, X_1, ..., X_{N-1}. In the beginning of the previous section we learned that each such function has a unique polynomial representation. We prove the claim on degree by induction. Before the first query, the state of the circuit is clearly a superposition of states (7.3) with constant coefficients. Assume, then, that after k queries and after applying operator U_k, the circuit state is a superposition

$$\sum_{\boldsymbol{x}\in\mathbb{F}_2^n} \sum_{b\in\mathbb{F}_2} \sum_{\boldsymbol{w}\in\mathbb{F}_2^m} P_{(\boldsymbol{x},b,\boldsymbol{w})} \, |\boldsymbol{x}\rangle \, |b\rangle \, |\boldsymbol{w}\rangle \, , \tag{7.6}$$

where each $P_{(\boldsymbol{x},b,\boldsymbol{w})}$ is a polynomial of variables X_0, X_1, ..., X_{N-1} having a degree of at most k. Then $k+1$th query will translate $|\boldsymbol{x}\rangle|0\rangle|\boldsymbol{w}\rangle$ into $|\boldsymbol{x}\rangle|1\rangle|\boldsymbol{w}\rangle$ and vice versa if $X_{\boldsymbol{x}} = f(\boldsymbol{x}) = 1$, but this translation will not happen if $X_{\boldsymbol{x}} = f(\boldsymbol{x}) = 0$ (again, we interprete $\boldsymbol{x}$ as a binary representation of an integer in $[0, N-1]$). This is to say that a partial sum

$$P_{(\boldsymbol{x},0,\boldsymbol{w})}|\boldsymbol{x}\rangle|0\rangle|\boldsymbol{w}\rangle + P_{(\boldsymbol{x},1,\boldsymbol{w})}|\boldsymbol{x}\rangle|1\rangle|\boldsymbol{w}\rangle \tag{7.7}$$

in (7.6) will transform into

$$\begin{aligned} &\left(X_{\boldsymbol{x}}P_{(\boldsymbol{x},1,\boldsymbol{w})} + (1-X_{\boldsymbol{x}})P_{(\boldsymbol{x},0,\boldsymbol{w})}\right)|\boldsymbol{x}\rangle|0\rangle|\boldsymbol{w}\rangle \\ +&\left(X_{\boldsymbol{x}}P_{(\boldsymbol{x},0,\boldsymbol{w})} + (1-X_{\boldsymbol{x}})P_{(\boldsymbol{x},1,\boldsymbol{w})}\right)|\boldsymbol{x}\rangle|1\rangle|\boldsymbol{w}\rangle . \end{aligned} \tag{7.8}$$

Using the assumption that $\deg P_{(\boldsymbol{x},b,\boldsymbol{w})} \le k$, (7.8) directly tells us that, after $k+1$ queries to Q_f, the coefficients are polynomials of degree at most $k+1$. Operator U_{k+1} produces a linear combination of these coefficient functions, which cannot increase the representation degrees. □

Remark 7.3.1. The transformation from (7.7) to (7.8) caused by Q_f is, of course, exactly the same as (7.4), only it is expressed in a different basis. By utilizing basis (7.5), Farhi & al. [37] have, independently from [6], derived a lower bound for the number of quantum queries needed to approximate PARITY.

Remark 7.3.2. Notice that the proof of Lemma 7.3.1 does not utilize the fact that each U_i is *unitary*; but rather, only the linearity is needed. In fact, if we could use nonlinear operators as well, reader may easily verify that we could have much faster growth in the representation degree of coefficients.

The following theorem from [6] finally connects quantum circuits and polynomial representations.

Theorem 7.3.1. *Let $N = 2^n$, $f : \mathbb{F}_2^n \to \mathbb{F}_2$ an arbitrary blackbox function and Q a quantum circuits that computes a Boolean function B on N variables $X_0 = f(0)$, $X_1 = f(1)$, ..., $X_{N-1} = f(N-1)$.*

1. *If Q computes B with a probability of 1, using T queries to f, then $T \ge \deg(B)/2$.*
2. *If Q computes B with a correctness probability of at least $\frac{2}{3}$ using T queries to f, then $T \ge \widetilde{\deg}(B)/2$.*

Proof. Without loss of generality, we may assume that Q gives the value of B by setting the right-most qubit to $B(X_0, X_1, \ldots, X_{N-1})$, which is then observed. By Lemma 7.3.1, the final state of Q is a superposition

$$\sum_{\boldsymbol{x}\in\mathbb{F}_2^n}\sum_{b\in\mathbb{F}_2}\sum_{\boldsymbol{w}\in\mathbb{F}_2^m} P_{(\boldsymbol{x},b,\boldsymbol{w})}|\boldsymbol{x}\rangle|b\rangle|\boldsymbol{w}\rangle ,$$

where each coefficient $P_{(\boldsymbol{x},b,\boldsymbol{w})}$ is a function of $X_0, \ldots, X_{N-1}$, which has a polynomial representation with a degree of at most T. The probability of seeing 1 on the right-most qubit is given by

$$P(X_0, \ldots, X_{N-1}) = \sum \left| P_{(\boldsymbol{x},b,\boldsymbol{w})} \right|^2, \tag{7.9}$$

where the sum is taken over all those states where the right-most bit of $\boldsymbol{w}$ is 1. By considering separately the real and imaginary parts, we see that $P(X_0, \ldots, X_{N-1})$, in fact, can be represented as a polynomial of variables $X_0, \ldots, X_{N-1}$, with real coefficients and degree at most $2T$.

If $\mathcal{Q}$ computes B exactly, then (7.9) is 1 if and only if B is 1, and, therefore, $P(X_0, \ldots, X_{N-1}) = B$. Thus, $2T \geq \deg(P) = \deg(B)$, and (1) follows. If $\mathcal{Q}$ computes B with a probability of at least $\frac{2}{3}$, then (7.9) is at most $\frac{1}{3}$ apart from 1 when B is 1, and similarly, (7.9) is at most $\frac{1}{3}$ apart from 0 when B is 0. This means that (7.9) is a polynomial approximating B, so $2T \geq \deg(P) \geq \widetilde{\deg}(B)$, and (2) follows. □

7.3.2 Some Examples

We can apply the main theorem (Theorem 7.3.1) of the previous section by taking $B = \text{OR}$ to arrive at the following theorem.

Theorem 7.3.2. *If $\mathcal{Q}$ is a quantum algorithm that makes T queries to a blackbox function f and decides with a correctness probability of at least $\frac{2}{3}$ if there is an element $\boldsymbol{x} \in \mathbb{F}_2^n$ such that $f(\boldsymbol{x}) = 1$, then $T \geq c\sqrt{2^n}$, where c is a constant.*

Proof. To decide if $f(\boldsymbol{x}) = 1$ for some $\boldsymbol{x} \in \mathbb{F}_2^n$ is to compute function OR on values $f(0)$, $f(1)$, …, $f(2^n - 1)$. According to Theorem 7.2.1, OR on $N = 2^n$ variables has approximation degree of $\Omega(\sqrt{N})$, so, by Theorem 7.3.1, computing OR on $N = 2^n$ variables (with a bounded probability of error) requires $\Omega(\sqrt{N}) = \Omega(\sqrt{2^n})$ queries to f. □

The above theorem shows that the method of using Grover's search algorithm to decide whether a solution for f exists, is optimal up to a constant factor. Similar results can be derived for other functions whose representation degree is known.

Theorem 7.3.3. *To decide if a blackbox function $f : \mathbb{F}_2^n \to \mathbb{F}_2$ has an even number of solutions, it requires $\Omega(2^n)$ queries to f for any quantum circuit.*

Proof. Take $B = \text{PARITY}$, apply Paturi's theorem (Theorem 7.2.1), and use reasoning similar to the above proof. □

Remark 7.3.3. In [6], the authors also derive results analogous to the previous ones for so-called Las Vegas quantum circuits, which must always give the correct answer, but that can, sometimes, be "ignorant", i.e., give the answer "I do not know", with a probability of at most $\frac{1}{2}$. Moreover, in [6] it is shown that if a quantum circuit computes a Boolean function B on variables X_0, X_1, ..., X_{N-1} using T queries to f with a nonvanishing error probability, then there is a classical deterministic algorithm that computes B exactly and uses $O(T^6)$ queries to f. Moreover, if B is symmetric, then $O(T^6)$ can even be replaced with $O(T^2)$.

Remark 7.3.4. It can be shown that only a vanishing ratio of Boolean functions on N variables have representation degree lower than N and that the same holds true for the approximation degree [2]. Thus, for almost any B, it requires $\Omega(N)$ queries to f to compute B on $f(0)$, $f(1)$, ..., $f(N-1)$ even with a bounded error probability.

Remark 7.3.5. With a little effort, we can utilize the results represented in this section to oracle Turing Machine computation: for almost all oracles X, $\mathbf{NP}^X$ *is not* included in $\mathbf{BQP}^X$. However, this does not imply that **NP** is not included in **BQP**, but it offers some reason to believe so. On the other hand, blackbox functions are the most "unstructured" examples that one can have – in [29] (see also [21]) W. van Dam has an example on breaking the blackbox lower bound by replacing an arbitrary f with a more structured function.

Remark 7.3.6. The lower bound by polynomial degree is not tight: A. Ambainis has proved the existence of a Boolean function that has degree M, but quantum query complexity $\Omega(M^{1.321\ldots})$. For the construction of the function and the lower bound method using the *quantum adversary technique*, see [3].

8. Appendix A: Quantum Physics

8.1 A Brief History of Quantum Theory

We may say that quantum mechanics was born in the beginning of the twentieth century when experiments on atoms, molecules and radiation physics were not explained by classical physics. As an example, we mention the quantum hypothesis by Max Planck in 1900[1] in connection with the radiation of a *black body*.

A black body is a hypothetical object that is able to emit and absorb electromagnetic radiation of all possible wavelengths. Although we cannot manufacture ideal black bodies, for experimental purposes, a small hole in an enclosure is a good approximation of a black body. Theoretical considerations revealed that the radiation intensity on different frequencies[2] should be the same for all black bodies. In the nineteenth century, there were two contradictory theories which described how the curve depicting this intensity should look: the first one was Wien's radiation law; the second one was Rayleigh-Jean's law of radiation. Both of them failed to predict the observed *frequency spectrum*, i.e., the distribution describing how the intensity of the radiation depends on the frequency.

The spectrum of the radiation became understandable when Planck published his radiation law [88]. An especially remarkable feature of Planck's work was that his law of radiation was derived under the assumption that electromagnetic radiation is emitted in discrete quanta whose energy E is proportional to the frequency ν:

$$E = h\nu. \tag{8.1}$$

The constant h was eventually uncovered by studying the experimental data. This famous constant is called *Planck's constant* and its value is approximately given by

$$h = 6.62608 \cdot 10^{-34} Js. \tag{8.2}$$

---

[1] M. Planck introduced his ideas in a lecture at the German Physical Society on December 14, 1900. As a general refrerence to early works on quantum physics, we mention [88].

[2] We could just as well talk about wavelength λ instead of the frequency ν; these quantities are related by formula $\nu\lambda = c$, where c is the vacuum speed of light.

Planck's quantum hypothesis contains a somewhat disturbing idea: since the days of Galileo and Newton, there had been a debate in the scientific community on whether light should be regarded as a particle flow or as waves. This discussion seemed to cease in the nineteenth century, mainly for the following reasons. First, in the beginning the nineteenth century, Young demonstrated by his famous two-slit experiment that light inherently has an undulatory nature. The second reason was that research at the end of the nineteenth century showed that light is, in fact, electromagnetic radiation, which is better described as waves. The research mentioned was mainly carried out by Maxwell and Hertz.

At least implicitly, Planck again called upon the old question about the nature of the light: the explanation of black body radiation, which finally agreed with observations in accordance with measurement precision, was derived under a hypothesis that light is made of discrete quanta.

Later, in 1905, Albert Einstein explained the *photoelectric effect*, basing the explanation on Planck's quantum hypothesis [88]. The photoelectric effect means that negatively charged metal loses its electrical charge when it is exposed to electromagnetic radiation of a certain frequency. It was not understood why the photoelectric effect so fundamentally depended on the frequency of the radiation and not on the intensity. In fact, reducing the intensity will slow down the effect, but cutting down the frequency will completely stop the effect. Einstein pointed out that, under Planck's hypothesis (8.2), this dependency on frequency becomes quite natural: the radiation should be regarded as a stream of particles whose energy is proportional to the frequency. It should be then quite clear that increasing the frequency of the radiation will also increase the impact of the radiation quanta on the electrons. When the impact becomes large enough, the electrons fly off and the negative charge disappears.[3] Einstein also introduced the notion of the *photon*, a light quantum.[4]

Planck's radiation law and Einstein's explanation for photoelectric effect opened the door for a novel idea: electromagnetic radiation, that was traditionally treated as a wave movement, possesses some characteristic features of a particle flow.

The experimental energy spectrum of a hydrogen atom was explained by Niels Bohr [88] up to the measurement precision of 1912. The model concerning hydrogen atoms introduced by Bohr provided some evidence that electrons, which had been considered as particles, may have wave-like characteristics as well. More evidence of the wave-like characteristics of electrons

[3] It should be emphasized here that the explanation given by Einstein is represented here only as a historical remark. According to modern quantum physics, the photoelectric effect can also be explained by assuming the undulatory nature of light and the quantum mechanical nature of the matter.

[4] In 1921, Albert Einstein was given the Nobel prize for the explanation of the photoelectric effect. The theory of relativity was not mentioned as a reason for Nobel prize.

came to light in the interference experiment carried out by C. J. Davidsson and L. H. Germer in 1927.

Inspired by the previous results, Luis de Broglie introduced a general hypothesis in 1924: particles may also be described as waves[5] whose wavelength λ can be uncovered when the momentum is known:

$$\lambda = \frac{h}{p}, \tag{8.3}$$

where h is Planck's constant (8.2).

Other famous physicists who have greatly influenced the development of quantum physics in the form we know it today are W. Heisenberg, M. Born, P. Jordan, W. Pauli, and P. A. M. Dirac.

8.2 Mathematical Framework for Quantum Theory

In this section, we shall first introduce the formalism of quantum mechanics in a basic form based on *state vectors.* Later we will also study a more general formalism based on *self-adjoint operators*, and we will see that the state vector formalism can be produced as a restriction of the more general formalism. The advantage of the state vector formalism is that it is mathematically simpler than the more general one.

In connection with quantum computation, we are primarily interested in representing a finite set[6] by using a quantum mechanical system. Therefore, we will make another significant mathematical simplification by assuming that all the quantum systems handled in this chapter are *finite-dimensional,* unless explicitly stated otherwise.

In the introductory chapter, we stated that we can describe a probabilistic system (with a finite phase space) by using a probability distribution

$$p_1[\boldsymbol{x}_1] + \ldots + p_n[\boldsymbol{x}_n] \tag{8.4}$$

over all the configurations $\boldsymbol{x}_i$. In the above mixed state, $p_1 + \ldots + p_n = 1$ and the system can be seen in state $\boldsymbol{x}_i$ with a probability of p_i. A quantum mechanical counterpart of this system is called an *n-level quantum system.* To describe such a system, we choose a basis $|\boldsymbol{x}_1\rangle, \ldots, |\boldsymbol{x}_n\rangle$ of an n-dimensional Hilbert space H_n, and a general state of an n-level quantum system is described by a vector

$$\alpha_1 |\boldsymbol{x}_1\rangle + \ldots + \alpha_n |\boldsymbol{x}_n\rangle, \tag{8.5}$$

[5] "There are not two worlds, one of light and waves, one of matter and corpuscles. There is only a single universe. Some of its properties can be accounted for by wave theory, and others by the corpuscular theory." (A citation of the lecture given by Louis de Broglie at the Nobel prize award ceremony in 1929.)

[6] In formal language theory, we also call a finite set representing information an *alphabet.*

where each α_i is a complex number called the *amplitude* of $\boldsymbol{x}_i$ and $|\alpha_1|^2 + \ldots + |\alpha_n|^2 = 1$. The basis $|\boldsymbol{x}_1\rangle, \ldots, |\boldsymbol{x}_n\rangle$ refers to an *observable* that can have some values, but for a moment we will ignore the values and say that the system may have properties $\boldsymbol{x}_1, \ldots, \boldsymbol{x}_n$ (with respect to the basis chosen). The probability that the system is seen to have property $\boldsymbol{x}_i$ is $|\alpha_i|^2$. Representation (8.5) has some properties that (8.4) does not have. For example, the time evolution of (8.4) sends each basis vector again into a combination of all the basis vectors with non-negative coefficients that sum up to 1. On the other hand, (8.5) may evolve in a very different way, as the basis states can also cancel each other.

Example 8.2.1. Consider a quantum physical description of two-state system having states 0 and 1. Assume also that the time evolution of the system (during a fixed time interval) is given by

$$
\begin{aligned}
|0\rangle &\mapsto \frac{1}{\sqrt{2}}|0\rangle + \frac{1}{\sqrt{2}}|1\rangle, \\
|1\rangle &\mapsto \frac{1}{\sqrt{2}}|0\rangle - \frac{1}{\sqrt{2}}|1\rangle.
\end{aligned}
$$

If the system starts in state $|0\rangle$ or $|1\rangle$ and undergoes the time evolution, the probability of observing 0 or 1 will be $\frac{1}{2}$ in both cases. On the other hand, if the system starts at state $|0\rangle$ and undergoes the time evolution *twice*, the state will be

$$
\begin{aligned}
&\frac{1}{\sqrt{2}}(\frac{1}{\sqrt{2}}|0\rangle + \frac{1}{\sqrt{2}}|1\rangle) + \frac{1}{\sqrt{2}}(\frac{1}{\sqrt{2}}|0\rangle - \frac{1}{\sqrt{2}}|1\rangle) \\
&= \frac{1}{2}|0\rangle + \frac{1}{2}|1\rangle + \frac{1}{2}|0\rangle - \frac{1}{2}|1\rangle = |0\rangle,
\end{aligned}
$$

and the probability of observing 0 becomes 1 again. The effect that the amplitudes of $|1\rangle$ cancel each other, is called *destructive interference* and the effect that the coefficients of $|0\rangle$ amplify each other is called *constructive interference*. Destructive interference cannot occur in the evolution of the probability distribution (8.4) since all the coefficients are always non-negative real numbers. A probabilistic counterpart to the quantum time evolution would be

$$
\begin{aligned}
[0] &\mapsto \frac{1}{2}[0] + \frac{1}{2}[1], \\
[1] &\mapsto \frac{1}{2}[0] + \frac{1}{2}[1],
\end{aligned}
$$

but the double time evolution beginning with state [0] would give state $\frac{1}{2}[0] + \frac{1}{2}[1]$ as the outcome.

Amplitude distribution (8.5) can naturally be interpreted as a unit-length vector of a Hilbert space. Therefore, the following sections are devoted to the study of Hilbert spaces (see also Section 9.3).

8.2.1 Hilbert Spaces

A finite-dimensional Hilbert space H is a complete (completeness will be defined later) vector space over complex numbers which is equipped with an inner product $H \times H \to \mathbb{C}$, $(\boldsymbol{x}, \boldsymbol{y}) \mapsto \langle \boldsymbol{x} \mid \boldsymbol{y} \rangle$. All n-dimensional Hilbert spaces are isomorphic, and we can, therefore, denote any such space by H_n. An inner product is required to satisfy the following axioms for all $\boldsymbol{x}$, $\boldsymbol{y}$, $\boldsymbol{z} \in H$ and c_1, $c_2 \in \mathbb{C}$:

1. $\langle \boldsymbol{x} \mid \boldsymbol{y} \rangle = \langle \boldsymbol{y} \mid \boldsymbol{x} \rangle^*$.
2. $\langle \boldsymbol{x} \mid \boldsymbol{x} \rangle \geq 0$ and $\langle \boldsymbol{x} \mid \boldsymbol{x} \rangle = 0$ if and only if $\boldsymbol{x} = \boldsymbol{0}$.
3. $\langle \boldsymbol{x} \mid c_1 \boldsymbol{y} + c_2 \boldsymbol{z} \rangle = c_1 \langle \boldsymbol{x} \mid \boldsymbol{y} \rangle + c_2 \langle \boldsymbol{x} \mid \boldsymbol{z} \rangle$.

If $E = \{\boldsymbol{e}_1, \ldots, \boldsymbol{e}_n\}$ is an orthonormal basis of H, each vector $\boldsymbol{x} \in H$ can be represented as $\boldsymbol{x} = x_1 \boldsymbol{e}_1 + \ldots x_n \boldsymbol{e}_n$. With a fixed basis E, we also use *coordinate representation* $\boldsymbol{x} = (x_1, \ldots, x_n)$. Using this representation one can easily see that an orthonormal basis induces an inner product by

$$\langle \boldsymbol{x} \mid \boldsymbol{y} \rangle = x_1^* y_1 + \ldots x_n^* y_n. \tag{8.6}$$

On the other hand, each inner product is induced by some orthonormal basis as in (8.6). For, if $\langle \cdot \mid \cdot \rangle$ stands for an arbitrary inner product, we may use the Gram-Schmidt process (see Section 9.3) to find a basis $\{\boldsymbol{b}_1, \ldots, \boldsymbol{b}_n\}$ orthonormal with respect to $\langle \cdot \mid \cdot \rangle$. Then

$$\begin{aligned} \langle \boldsymbol{x} \mid \boldsymbol{y} \rangle &= \langle x_1 \boldsymbol{b}_1 + \ldots + x_n \boldsymbol{b}_n \mid y_1 \boldsymbol{b}_1 + \ldots y_n \boldsymbol{b}_n \rangle \\ &= x_1^* y_1 + \ldots + x_n^* y_n \end{aligned}$$

The inner product induces a *vector norm* in a very natural way:

$$||\boldsymbol{x}|| = \sqrt{\langle \boldsymbol{x} \mid \boldsymbol{x} \rangle}.$$

Informally speaking, the *completeness* of the vector space H means that there are enough vectors in H, i.e., at least one for each limit process:

Definition 8.2.1. *A vector space H is complete, if for each vector sequence $\boldsymbol{x}_i$ such that*

$$\lim_{m,n \to \infty} ||\boldsymbol{x}_m - \boldsymbol{x}_n|| = 0,$$

there exists a vector $\boldsymbol{x} \in H$ such that

$$\lim_{n \to \infty} ||\boldsymbol{x}_n - \boldsymbol{x}|| = 0.$$

An important geometric property of Hilbert spaces is that each subspace $W \in H$ which is also a Hilbert space,[7] has an orthogonal complement. The proofs of the following lemmata are left as exercises.

[7] In finite-dimensional Hilbert spaces, all the subspaces are also Hilbert spaces.

Lemma 8.2.1. *Let W be a subspace of H. Then the set of vectors*

$$W^\perp = \{\boldsymbol{y} \in H \mid \langle \boldsymbol{y} \mid \boldsymbol{x} \rangle = 0 \text{ whenever } \boldsymbol{x} \in H.\}$$

is also a subspace of H, which is called the orthogonal complement of W.

Lemma 8.2.2. *If W is a subspace of finite-dimensional Hilbert space H, then $H = W \oplus W^\perp$.*

8.2.2 Operators

The modern description of quantum mechanics is profoundly based on linear mappings. In the following sections, we represent the features of the linear mappings that are most essential for quantum mechanics. Because we concentrate mainly on finite-level quantum systems, the vector spaces that are treated hereafter will be assumed to have finite dimensions, unless explicitly stated otherwise.

Let us begin with some terminology: a linear mapping $H \to H$ is called an operator. The set of operators on H is denoted by $L(H)$. For an operator T, we define the *norm* of the operator by

$$||T|| = \sup_{||\boldsymbol{x}||=1} ||T\boldsymbol{x}||.$$

A nonzero vector $\boldsymbol{x} \in H$ is an *eigenvector* of T belonging to *eigenvalue* $\lambda \in \mathbb{C}$, if $T\boldsymbol{x} = \lambda\boldsymbol{x}$.

The set of operators also forms a vector space, if the sum and scalar multiplication is defined in natural way: Let S and $T \in L(H)$ and $a \in \mathbb{C}$. Then $S + T$ and aS are operators in $L(H)$ defined as

$$(S + T)\boldsymbol{x} = S\boldsymbol{x} + T\boldsymbol{x}$$

and

$$(aS)\boldsymbol{x} = a(S\boldsymbol{x})$$

for each $\boldsymbol{x} \in H$.

Definition 8.2.2. *For any operator $T : H \to H$, the adjoint operator T^* is defined by the requirement*

$$\langle \boldsymbol{x} \mid T\boldsymbol{y} \rangle = \langle T^*\boldsymbol{x} \mid \boldsymbol{y} \rangle$$

for all $\boldsymbol{x}$, $\boldsymbol{y} \in H$.

Remark 8.2.1. With a fixed basis $\{\boldsymbol{e}_1, \ldots, \boldsymbol{e}_n\}$ of H_n, any operator T can be represented as an $n \times n$ matrix over the field of complex numbers. It is not difficult to see that the matrix representing the adjoint operator T^* is the transposed complex conjugate of the matrix representing T.

If $\{\boldsymbol{x}_1, \ldots, \boldsymbol{x}_n\}$ and $\{\boldsymbol{y}_1, \ldots, \boldsymbol{y}_n\}$ are orthonormal bases of H_n, then it can be shown that

$$\sum_{i=1}^{n} \langle \boldsymbol{x}_i \mid T\boldsymbol{x}_i \rangle = \sum_{i=1}^{n} \langle \boldsymbol{y}_i \mid T\boldsymbol{y}_i \rangle; \tag{8.7}$$

see Exercise 1.

Definition 8.2.3. *For an orthonormal basis $\{\boldsymbol{x}_1, \ldots, \boldsymbol{x}_n\}$, quantity*

$$\mathrm{Tr}(T) = \sum_{i=1}^{n} \langle \boldsymbol{x}_i \mid T\boldsymbol{x}_i \rangle$$

is called the trace of operator T.

By (8.7), the notion of trace is well-defined. Moreover, it is clear that the trace is linear. Notice also that in the matrix representation of T, the trace is the sum of the diagonal elements.

Definition 8.2.4. *An operator T is self-adjoint if $T^* = T$. An operator T is unitary if $T^* = T^{-1}$.*

The following simple lemmata will be used frequently.

Lemma 8.2.3. *A self-adjoint operator has real eigenvalues.*

Proof. If $A\boldsymbol{x} = \lambda\boldsymbol{x}$, then

$$\lambda^* \langle \boldsymbol{x} \mid \boldsymbol{x} \rangle = \langle \lambda\boldsymbol{x} \mid \boldsymbol{x} \rangle = \langle A\boldsymbol{x} \mid \boldsymbol{x} \rangle = \langle \boldsymbol{x} \mid A\boldsymbol{x} \rangle = \lambda \langle \boldsymbol{x} \mid \boldsymbol{x} \rangle.$$

Since $\boldsymbol{x} \neq \boldsymbol{0}$ as an eigenvector, if follows that $\lambda^* = \lambda$. □

Lemma 8.2.4. *The eigenvectors of a self-adjoint operator belonging to distinct eigenvalues are orthogonal.*

Proof. Assume that $\lambda \neq \lambda'$, $A\boldsymbol{x} = \lambda\boldsymbol{x}$, and $A\boldsymbol{x}' = \lambda'\boldsymbol{x}'$. Since λ and λ' are real by the previous lemma,

$$\lambda' \langle \boldsymbol{x}' \mid \boldsymbol{x} \rangle = \langle A\boldsymbol{x}' \mid \boldsymbol{x} \rangle = \langle \boldsymbol{x}' \mid A\boldsymbol{x} \rangle = \lambda \langle \boldsymbol{x}' \mid \boldsymbol{x} \rangle,$$

and therefore $\langle \boldsymbol{x}' \mid \boldsymbol{x} \rangle = 0$. □

Definition 8.2.5. *A self-adjoint operator T is positive if $\langle \boldsymbol{x} \mid T\boldsymbol{x} \rangle \geq 0$ for each $\boldsymbol{x} \in H$.*

Remark 8.2.2. A partial order in the set of operators can be introduced by defining $T \geq S$ if and only if $T - S$ is positive.

If W is a subspace of H and $H = W \oplus W^\perp$, then each vector $\boldsymbol{x} \in H$ can be uniquely represented as $\boldsymbol{x}_W + \boldsymbol{x}_{W^\perp}$, where $\boldsymbol{x}_W \in W$ and $\boldsymbol{x}_{W^\perp} \in W^\perp$. It is quite easy to verify that the mapping P_W defined by $P_W(\boldsymbol{x}_W + \boldsymbol{x}_{W^\perp}) = \boldsymbol{x}_W$ is a self-adjoint linear mapping, called the *projection* onto subspace W. Clearly, $P_W^2 = P_W$. On the other hand, it can be shown that each self-adjoint P such that $P^2 = P$ is a projection onto some subspace of H (Exercise 2). The set of projections in $L(H)$ is denoted by $P(H)$. Notice that a projection can never increase the norm:

$$||P_W\boldsymbol{x}||^2 = ||\boldsymbol{x}_W||^2 \leq ||\boldsymbol{x}_W||^2 + ||\boldsymbol{x}_{W^T}||^2 = ||\boldsymbol{x}_W + \boldsymbol{x}_{W^T}||^2 = ||\boldsymbol{x}||^2.$$

The second-last equality, which is called *Pythagoras' theorem*, follows from the fact that $\boldsymbol{x}_W$ and $\boldsymbol{x}_{W^T}$ are orthogonal. Projections play an important role in the theory of self-adjoint operators, and so we will study them in more detail in the next section.

Since unitary operators also play a great role in quantum theory, we will study some of their properties.

Lemma 8.2.5. *Unitary operator $U : H \to H$ preserves the inner products, that is, $\langle U\boldsymbol{x} \mid U\boldsymbol{y}\rangle = \langle \boldsymbol{x} \mid \boldsymbol{y}\rangle$.*

Proof. This follows directly from the definition of the adjoint operator and unitarity:

$$\langle U\boldsymbol{x} \mid U\boldsymbol{y}\rangle = \langle U^*U\boldsymbol{x} \mid \boldsymbol{y}\rangle = \langle U^{-1}U\boldsymbol{x} \mid \boldsymbol{y}\rangle = \langle \boldsymbol{x} \mid \boldsymbol{y}\rangle.$$

□

Corollary 8.2.1. *Unitary operators preserve norms, i.e., $||U\boldsymbol{x}|| = ||\boldsymbol{x}||$.*

A statement converse to the previous lemma also holds:

Lemma 8.2.6. *Operators $U : H \to H$ that preserve inner products are unitary.*

Proof. Assume that $\langle U\boldsymbol{x} \mid U\boldsymbol{y}\rangle = \langle \boldsymbol{x} \mid \boldsymbol{y}\rangle$ for each $\boldsymbol{x}, \boldsymbol{y} \in H$. Then, especially $||U\boldsymbol{x}|| = ||\boldsymbol{x}||$ for each $\boldsymbol{x} \in H$ which implies that U is injective. Since H was assumed finite-dimensional, U is surjective, too. Therefore U^{-1} exists, and

$$\langle \boldsymbol{x} \mid U^{-1}\boldsymbol{y}\rangle = \langle U\boldsymbol{x} \mid UU^{-1}\boldsymbol{y}\rangle = \langle U\boldsymbol{x} \mid \boldsymbol{y}\rangle,$$

which means that $U^* = U^{-1}$, i.e., U is unitary. □

Lemma 8.2.7. *Operators $U : H \to H$ that preserve norms are unitary.*

Proof. The *polarization equation*

$$\langle \boldsymbol{x} \mid \boldsymbol{y}\rangle = \frac{1}{4}\sum_{k=0}^{3} i^k \langle \boldsymbol{y} + i^k\boldsymbol{x} \mid \boldsymbol{y} + i^k\boldsymbol{x}\rangle \qquad (8.8)$$

is easy to verify by straightforward calculation (Exercise 4). From (8.8), it follows directly that operators which preserve norms, also preserve inner products. Now, the claim follows by Lemma 8.2.6. □

We can even strengthen Lemma 8.2.6 a little bit:

Lemma 8.2.8. *Let* $\{\boldsymbol{x}_1, \ldots, \boldsymbol{x}_k\}$ *and* $\{\boldsymbol{y}_1, \ldots, \boldsymbol{y}_k\}$ *be two sets of vectors in* H*. If* $\langle \boldsymbol{x}_i \mid \boldsymbol{x}_j \rangle = \langle \boldsymbol{y}_i \mid \boldsymbol{y}_j \rangle$ *for each* i *and* j*; then there is a unitary mapping* $U : H \to H$ *such that* $\boldsymbol{y}_i = U\boldsymbol{x}_i$*.*

Proof. Let W be the subspace of H generated by vectors $\boldsymbol{x}_1, \ldots, \boldsymbol{x}_k$. There exists a subset of $\{\boldsymbol{x}_1, \ldots, \boldsymbol{x}_k\}$ which forms a basis of W. Without loss of generality, we may assume that this subset is $\{\boldsymbol{x}_1, \ldots, \boldsymbol{x}_{k'}\}$ for some $k' \leq k$. Now we define a mapping $U : W \to H$ by $U\boldsymbol{x}_i = \boldsymbol{y}_i$ for each $i \in \{1, \ldots, k'\}$ and extend this into a linear mapping in the only possible way.

We will first show that $\boldsymbol{y}_1, \ldots, \boldsymbol{y}_{k'}$ is a basis of $\mathrm{Im}(U)$. Clearly those vectors generate $\mathrm{Im}(U)$, so it remains to show that they are linearly independent. For that purpose, we assume that

$$a_1\boldsymbol{y}_1 + \ldots + a_{k'}\boldsymbol{y}_{k'} = \boldsymbol{0} \tag{8.9}$$

for some coefficients $a_1, \ldots, a_{k'}$. For any $i \in \{1, \ldots, k'\}$, we compute the inner product of (8.9) by $\boldsymbol{y}_i$ thus getting

$$a_1\langle \boldsymbol{y}_i \mid \boldsymbol{y}_1 \rangle + \ldots + a_k\langle \boldsymbol{y}_i \mid \boldsymbol{y}_{k'} \rangle = 0. \tag{8.10}$$

On the other hand, by the assumption, (8.10) can be written as

$$a_1\langle \boldsymbol{x}_i \mid \boldsymbol{x}_1 \rangle + \ldots + a_k\langle \boldsymbol{x}_i \mid \boldsymbol{x}_{k'} \rangle = 0. \tag{8.11}$$

Equation (8.11) states that

$$\langle \boldsymbol{x}_i \mid a_1\boldsymbol{x}_1 + \ldots + a_{k'}\boldsymbol{x}_{k'} \rangle = 0$$

for each $i \in \{1, \ldots, k'\}$, which implies that the vector

$$a_1\boldsymbol{x}_1 + \ldots + a_{k'}\boldsymbol{x}_{k'}. \tag{8.12}$$

is orthogonal to all vectors in space W. Since vector (8.12) itself belongs to W, (8.12) must be the zero vector. Therefore $a_1 = \ldots = a_{k'} = 0$, and it follows that $\boldsymbol{y}_1, \ldots, \boldsymbol{y}_{k'}$ forms a basis of $\mathrm{Im}(U)$.

It follows directly that $U : W \to H$ is injective: the equation

$$U(a_1\boldsymbol{x}_1 + \ldots + a_{k'}\boldsymbol{x}_{k'}) = \boldsymbol{0}$$

means that

$$a_1\boldsymbol{y}_1 + \ldots + a_{k'}\boldsymbol{y}_{k'} = \boldsymbol{0},$$

and this implies that $a_1 = \ldots = a_{k'} = 0$.

Now, having a bijection $U : W \to \mathrm{Im}(U)$, we extend it to the whole space H as follows: Let $\{\boldsymbol{z}_1, \ldots, \boldsymbol{z}_r\}$ be an orthonormal basis of $W^{\perp}$ and $\{\boldsymbol{z}'_1, \ldots,$

$\boldsymbol{z}'_r\}$ an orthonormal basis of $\mathrm{Im}(W)^\perp$. Defining the extension of U so that $U(\boldsymbol{z}_i) = \boldsymbol{z}'_i$ for each $i \in \{1, \ldots, r\}$, mapping $U : H \to H$ obtained in this way is clearly bijective. To prove that U is unitary, by Lemma 8.2.7, it remains to show that U preserves the inner products. For this purpose, it is sufficient (by linearity) to show that U preserves all the inner products of the basis vectors $\{\boldsymbol{x}_1, \ldots, \boldsymbol{x}_{k'}, \boldsymbol{z}_1, \ldots, \boldsymbol{z}_r\}$. This is quite easy: $\langle U\boldsymbol{x}_i \mid U\boldsymbol{x}_j\rangle = \langle \boldsymbol{y}_i \mid \boldsymbol{y}_j\rangle = \langle \boldsymbol{x}_i \mid \boldsymbol{x}_j\rangle$ for each $i, j \in \{1, \ldots, k'\}$. Second, $\langle U\boldsymbol{x}_i \mid U\boldsymbol{z}_j\rangle = \langle \boldsymbol{y}_i \mid \boldsymbol{z}'_i\rangle = 0 = \langle \boldsymbol{x}_i \mid \boldsymbol{z}_i\rangle$, and finally, $\langle U\boldsymbol{z}_i \mid U\boldsymbol{z}_j\rangle = \langle \boldsymbol{z}'_i \mid \boldsymbol{z}'_j\rangle = \delta_{ij} = \langle \boldsymbol{z}_i \mid \boldsymbol{z}_j\rangle$.

Finally, we have to show that $\boldsymbol{y}_i = U\boldsymbol{x}_i$ for each $i \in \{1, \ldots, k\}$. If $i \leq k'$, there is nothing to prove, so we may assume that $i > k'$. Now that $\boldsymbol{x}_1, \ldots, \boldsymbol{x}_{k'}$ form a basis of W, each vector $\boldsymbol{x}_i \in W$ can be uniquely written as

$$\boldsymbol{x}_i = c_1^{(i)}\boldsymbol{x}_1 + \ldots + c_{k'}^{(i)}\boldsymbol{x}_{k'}.$$

It follows directly that

$$\langle \boldsymbol{x}_j \mid \boldsymbol{x}_i\rangle = c_1^{(i)}\langle \boldsymbol{x}_j \mid \boldsymbol{x}_1\rangle + \ldots + c_{k'}^{(i)}\langle \boldsymbol{x}_j \mid \boldsymbol{x}_{k'}\rangle. \tag{8.13}$$

By the assumption, (8.13) can be rewritten as

$$\langle \boldsymbol{y}_j \mid \boldsymbol{y}_i\rangle = c_1^{(i)}\langle \boldsymbol{y}_j \mid \boldsymbol{y}_1\rangle + \ldots + c_{k'}^{(i)}\langle \boldsymbol{y}_j \mid \boldsymbol{y}_{k'}\rangle,$$

or as

$$\langle \boldsymbol{y}_j \mid \boldsymbol{y}_i\rangle = \langle \boldsymbol{y}_j \mid c_1^{(i)}\boldsymbol{y}_1 + \ldots + c_{k'}^{(i)}\boldsymbol{y}_{k'}\rangle. \tag{8.14}$$

But 8.14 says that

$$\langle \boldsymbol{y}_j \mid \boldsymbol{y}_i - (c_1^{(i)}\boldsymbol{y}_1 + \ldots + c_{k'}^{(i)}\boldsymbol{y}_{k'})\rangle = 0$$

for each $\boldsymbol{y}_j$. Now we can conclude as earlier: vector $\boldsymbol{y}_i - (c_1^{(i)}\boldsymbol{y}_1 + \ldots + c_{k'}^{(i)}\boldsymbol{y}_{k'})$ is orthogonal to the subspace generated by vectors $\boldsymbol{y}_1, \ldots, \boldsymbol{y}_k$. Because it belongs to that space, we must conclude that it is the zero vector. Therefore,

$$\boldsymbol{y}_i = c_1^{(i)}\boldsymbol{y}_1 + \ldots + c_{k'}^{(i)}\boldsymbol{y}_{k'} = c_1^{(i)}U\boldsymbol{x}_1 + \ldots + c_{k'}^{(i)}U\boldsymbol{x}_{k'} = U\boldsymbol{x}_i.$$

□

8.2.3 Spectral Representation of Self-Adjoint Operators

For any vectors $\boldsymbol{x}$ and $\boldsymbol{y} \in H_n$, we define a linear mapping $|\boldsymbol{x}\rangle\langle\boldsymbol{y}|: H_n \to H_n$ by

$$|\boldsymbol{x}\rangle\langle\boldsymbol{y}|\, \boldsymbol{z} = \langle \boldsymbol{y} \mid \boldsymbol{z}\rangle\boldsymbol{x}.$$

It is plain to see that, if $||\boldsymbol{x}|| = 1$, then $|\,\boldsymbol{x}\rangle\langle\boldsymbol{x}\,|$ is the projection onto the one-dimensional subspace generated by $\boldsymbol{x}$.

Remark 8.2.3. If A and $B \in L(H_n)$, then

$$|\,A\boldsymbol{x}\rangle\langle B\boldsymbol{y}\,|\,\boldsymbol{z} = \langle B\boldsymbol{y} \mid \boldsymbol{z}\rangle A\boldsymbol{x} = A\langle \boldsymbol{y} \mid B^*\boldsymbol{z}\rangle\boldsymbol{x} = A\,|\,\boldsymbol{x}\rangle\langle\boldsymbol{y}\,|\,B^*\boldsymbol{z}.$$

Since this holds for each $\boldsymbol{z} \in H_n$, we have that $|\,A\boldsymbol{x}\rangle\langle B\boldsymbol{y}\,| = A\,|\,\boldsymbol{x}\rangle\langle\boldsymbol{y}\,|\,B^*$.

Remark 8.2.4. The adjoint operator of $|\,\boldsymbol{x}\rangle\langle\boldsymbol{y}\,|$ can be easily found: $\left(\,|\,\boldsymbol{x}\rangle\langle\boldsymbol{y}\,|\,\right)^* = |\,\boldsymbol{y}\rangle\langle\boldsymbol{x}\,|$.

Remark 8.2.5. If $\{\boldsymbol{x}_1, \ldots, \boldsymbol{x}_n\}$ is an orthonormal basis of H_n, then the matrix representation of mapping $|\,\boldsymbol{x}_i\rangle\langle\boldsymbol{x}_j\,|$ is simply the matrix having 0s elsewhere but 1 at the intersection of the ith row and the jth column. It follows directly that mappings $|\,\boldsymbol{x}_i\rangle\langle\boldsymbol{x}_j\,|$ generate the whole space $L(H_n)$. In fact, if $T \in L(H_n)$, then

$$T = \sum_{r=1}^{n}\sum_{s=1}^{n}\langle\boldsymbol{x}_r \mid T\boldsymbol{x}_s\rangle\,|\,\boldsymbol{x}_r\rangle\langle\boldsymbol{x}_s\,| \tag{8.15}$$

is the *matrix representation* of T in basis $\{\boldsymbol{x}_1, \ldots, \boldsymbol{x}_n\}$. It is easy to see that the mappings $|\,\boldsymbol{x}_i\rangle\langle\boldsymbol{x}_j\,|$ are linearly independent, and therefore the dimension of $L(H_n)$ is n^2.

To introduce the spectral decomposition of a self-adjoint operator T, we present the following well-known result:

Theorem 8.2.1. *Let $T : H_n \to H_n$ be a self-adjoint operator. There exists an orthonormal basis of H_n that consists of eigenvectors of T.*

Proof. The proof is made by induction on $n = \dim H_n$. One-dimensional H_1 is generated by some single vector $\boldsymbol{e}_1$ of unit length. Clearly $T\boldsymbol{e}_1 = \lambda\boldsymbol{e}_1$ for some $\lambda \in \mathbb{C}$, so the claim is obvious. If $n > 1$, let λ be an eigenvalue of T, and let W be the *eigenspace* of λ, that is, the subspace of H_n generated by the eigenvectors belonging to λ. Any linear combination of the eigenvectors belonging to λ is again an eigenvector belonging to λ, which implies that W is closed under T. But also, the orthogonal complement $W^\perp$ is closed under T: in fact, take any $\boldsymbol{x} \in W$ and $\boldsymbol{y} \in W^\perp$. Then,

$$\langle\boldsymbol{x} \mid T\boldsymbol{y}\rangle = \langle T\boldsymbol{x} \mid \boldsymbol{y}\rangle = \lambda^*\langle\boldsymbol{x} \mid \boldsymbol{y}\rangle = 0,$$

which means that $T\boldsymbol{y} \in W^\perp$ as well. Therefore, we may study the restrictions $T : W \to W$ and $T : W^\perp \to W^\perp$ and apply the induction hypothesis to find an orthonormal basis for W and $W^\perp$ consisting of eigenvectors of T. Since $H_n = W \oplus W^\perp$, the union of these bases satisfies the claim. □

We can now introduce *spectral representation*, a powerful tool in the study of self-adjoint mappings. Let $\boldsymbol{x}_1, \ldots, \boldsymbol{x}_n$ be orthonormal eigenvectors of a self-adjoint operator $T : H_n \to H_n$, and let $\lambda_1, \ldots, \lambda_n$ be the corresponding eigenvalues. Numbers λ_i are real by Theorem 8.2.1, but they are not necessarily distinct. The set of eigenvalues is called a *spectrum*, and it can be easily verified (Exercise 3) that

$$T = \lambda_1 \, |\boldsymbol{x}_1\rangle\langle\boldsymbol{x}_1| + \ldots + \lambda_n \, |\boldsymbol{x}_n\rangle\langle\boldsymbol{x}_n| \, . \tag{8.16}$$

Decomposition (8.16) is called a *spectral representation* of T. If T has multiple eigenvalues, we say that T is *degenerate*; otherwise, T is *nondegenerate*. The spectral representation (8.16) of a nondegenerate T is unique, as is easily verified. Notice that we do not claim that the vectors $\boldsymbol{x}_i$ are unique, only that the projections $|\, \boldsymbol{x}_i\rangle\langle\boldsymbol{x}_i \,|$ are. If T is degenerate, we can collect the multiple eigenvalues in (8.16) as common factors to obtain a representation

$$T = \lambda'_1 P_1 + \ldots + \lambda'_{n'} P_{n'}, \tag{8.17}$$

where $P_1, \ldots, P_{n'}$ are the projections onto the eigenspaces of $\lambda'_1, \ldots, \lambda'_{n'}$. It is easy to see that the spectral representation (8.17) is unique.[8]

Recall that all the eigenvectors belonging to distinct eigenvalues are orthogonal, which implies that, in representation (8.17), all the projections are projections to mutually orthogonal subspaces. Therefore, $P_i P_j = 0$ whenever $i \neq j$. It follows that, if p is a polynomial, then

$$p(T) = p(\lambda'_1)P_1 + \ldots + p(\lambda'_{n'})P_{n'}. \tag{8.18}$$

We also generalize (8.18): if $f : \mathbb{R} \to \mathbb{C}$ is *any* function and (8.17) is the spectral representation of T, we define

$$f(T) = f(\lambda'_1)P_1 + \ldots + f(\lambda'_{n'})P_{n'}. \tag{8.19}$$

Example 8.2.2. The *identity operator* $I \in L(H_n)$ is defined as $I\boldsymbol{x} = \boldsymbol{x}$ for each $\boldsymbol{x} \in H_n$. Operator I is trivially self-adjoint, and its only eigenvalue is 1. Moreover, the eigenspace of I is clearly the whole space H_n. To find a spectral representation of I, it therefore suffices to fix any orthonormal basis $\{\boldsymbol{x}_1, \ldots, \boldsymbol{x}_n$, and then we know that

$$I = |\boldsymbol{x}_1\rangle\langle\boldsymbol{x}_1| + \ldots + |\boldsymbol{x}_n\rangle\langle\boldsymbol{x}_n| \, . \tag{8.20}$$

Equation (8.20) could, of course, also be verified straightforwardly.

Example 8.2.3. Matrix

$$M_\neg = \begin{pmatrix} 0 & 1 \\ 1 & 0 \end{pmatrix}$$

defines a self-adjoint mapping in H_2, as is easily verified. Matrix $M_\neg$ is nondegenerate, having 1 and -1 as eigenvalues. The corresponding orthonormal

[8] Some authors call only the unique representation (8.17) a spectral representation, not representation (8.16).

eigenvectors can be chosen as $\boldsymbol{x}_1 = \frac{1}{\sqrt{2}}(1,1)^T$ and $\boldsymbol{x}_{-1} = \frac{1}{\sqrt{2}}(1,-1)^T$, for example. The matrix representations of both projections $|\,\boldsymbol{x}_{\pm 1}\rangle\langle\boldsymbol{x}_{\pm 1}\,|$ can be easily uncovered; they are

$$\begin{pmatrix} \frac{1}{2} & \frac{1}{2} \\ \frac{1}{2} & \frac{1}{2} \end{pmatrix} \text{ and } \begin{pmatrix} \frac{1}{2} & -\frac{1}{2} \\ -\frac{1}{2} & \frac{1}{2} \end{pmatrix},$$

respectively. Thus, the spectral representation of $M_\neg$ is given by

$$\begin{pmatrix} 0 & 1 \\ 1 & 0 \end{pmatrix} = 1 \cdot \begin{pmatrix} \frac{1}{2} & \frac{1}{2} \\ \frac{1}{2} & \frac{1}{2} \end{pmatrix} - 1 \cdot \begin{pmatrix} \frac{1}{2} & -\frac{1}{2} \\ -\frac{1}{2} & \frac{1}{2} \end{pmatrix}.$$

Now we can find the square root of $M_\neg$:

$$\sqrt{M_\neg} = \sqrt{1} \cdot \begin{pmatrix} \frac{1}{2} & \frac{1}{2} \\ \frac{1}{2} & \frac{1}{2} \end{pmatrix} + \sqrt{-1} \cdot \begin{pmatrix} \frac{1}{2} & -\frac{1}{2} \\ -\frac{1}{2} & \frac{1}{2} \end{pmatrix}.$$

Choosing $\sqrt{1} = 1$, $\sqrt{-1} = i$, we obtain the matrix of Example 2.1.1. By fixing the square roots above in all possible ways, we get four different matrices X that satisfy the condition $X^2 = M_\neg$.

8.2.4 Spectral Representation of Unitary Operators

Let us study a function e^{iT}, which is defined on a self-adjoint operator defined by spectral representation

$$T = \lambda_1 \,|\,\boldsymbol{x}_1\rangle\langle\boldsymbol{x}_1\,| + \ldots + \lambda_n \,|\,\boldsymbol{x}_n\rangle\langle\boldsymbol{x}_n\,|\,.$$

By definition,

$$\mathrm{e}^{iT} = \mathrm{e}^{i\lambda_1} \,|\,\boldsymbol{x}_1\rangle\langle\boldsymbol{x}_1\,| + \ldots + \mathrm{e}^{i\lambda_n} \,|\,\boldsymbol{x}_n\rangle\langle\boldsymbol{x}_n\,|,$$

and we notice that

$$(\mathrm{e}^{iT})^* = \mathrm{e}^{-i\lambda_1} \,|\,\boldsymbol{x}_1\rangle\langle\boldsymbol{x}_1\,| + \ldots + \mathrm{e}^{-i\lambda_n} \,|\,\boldsymbol{x}_n\rangle\langle\boldsymbol{x}_n\,| = (\mathrm{e}^{iT})^{-1},$$

which is to say that e^{iT} is unitary. We will now show that each unitary mapping can be represented as e^{iT}, where T is a self-adjoint operator. To do this, we first derive an auxiliary result which also has independent interest.

Lemma 8.2.9. *Self-adjoint operators A and B commute if and only if A and B share an orthonormal eigenvector basis.*

Proof. If $\{\boldsymbol{a}_1, \ldots, \boldsymbol{a}_n\}$ is a set of orthonormal eigenvectors of both A and B, then, by using the corresponding eigenvalues, we can write

$$A = \lambda_1 \,|\,\boldsymbol{a}_1\rangle\langle\boldsymbol{a}_1\,| + \ldots + \lambda_n \,|\,\boldsymbol{a}_n\rangle\langle\boldsymbol{a}_n\,|$$

and

$$B = \mu_1 \, |\boldsymbol{a}_1\rangle\langle\boldsymbol{a}_1| + \ldots + \mu_n \, |\boldsymbol{a}_n\rangle\langle\boldsymbol{a}_n| \, .$$

Hence,

$$AB = \lambda_1\mu_1 \, |\boldsymbol{a}_1\rangle\langle\boldsymbol{a}_1| + \ldots + \lambda_n\mu_n \, |\boldsymbol{a}_n\rangle\langle\boldsymbol{a}_n| = BA.$$

Assume, then, that $AB = BA$. Let $\lambda_1, \ldots, \lambda_h$ be all the *distinct* eigenvalues of A, and let $\boldsymbol{a}_1^{(k)}, \ldots, \boldsymbol{a}_{m_k}^{(k)}$ be orthonormal eigenvectors belonging to λ_k. For any $\boldsymbol{a}_i^{(k)}$, we have

$$AB\boldsymbol{a}_i^{(k)} = BA\boldsymbol{a}_i^{(k)} = B\lambda_k\boldsymbol{a}_i^{(k)} = \lambda_k B\boldsymbol{a}_i^{(k)},$$

i.e., $B\boldsymbol{a}_i^{(k)}$ is also an eigenvector of A belonging to eigenvalue λ_k. Therefore, the eigenspace of λ_k is closed under B. We denote this subspace by W_k. As a restriction of a self-adjoint operator, clearly $B : W_k \to W_k$ is also self-adjoint. Therefore, B has m_k eigenvectors $\boldsymbol{b}_1^{(k)}, \ldots, \boldsymbol{b}_{m_k}^{(k)}$ that form an orthonormal basis of W_k. By finding such a basis for each W_k, we obtain an orthonormal eigenvector system such that

- Each $\boldsymbol{b}_i^{(k)}$ is an eigenvector of B and A (for A, $\boldsymbol{b}_i^{(k)}$ is an eigenvector belonging to λ_k).
- If $i \neq j$, then $\boldsymbol{b}_i^{(k)}$ and $\boldsymbol{b}_j^{(k)}$ are orthogonal, since they belong to an orthonormal basis generating W_k.
- If $k \neq k'$, then $\boldsymbol{b}_i^{(k)}$ and $\boldsymbol{b}_j^{(k')}$ are orthogonal, since they are eigenvectors of A belonging to distinct eigenvalues λ_k and $\lambda_{k'}$.

Thus, the system which is obtained is an orthonormal basis of H_n but also a set of eigenvectors of both A and B. □

Now, let U be a unitary operator. We notice, first, that the eigenvalues of U have absolute values of 1. To verify this, let $\boldsymbol{x}$ be an eigenvector belonging to eigenvalue λ, i.e., $U\boldsymbol{x} = \lambda\boldsymbol{x}$. Then,

$$\langle\boldsymbol{x} \mid \boldsymbol{x}\rangle = \langle\boldsymbol{x} \mid U^*U\boldsymbol{x}\rangle = \langle U\boldsymbol{x} \mid U\boldsymbol{x}\rangle = \langle\lambda\boldsymbol{x} \mid \lambda\boldsymbol{x}\rangle = |\lambda|^2 \langle\boldsymbol{x} \mid \boldsymbol{x}\rangle,$$

and it follows that $|\lambda| = 1$.

We decompose U into "real and imaginary parts" by writing $U = A + iB$, where $A = \frac{1}{2}(U + U^*)$ and $B = \frac{1}{2i}(U - U^*)$. Note that A and B are now self-adjoint, commuting operators. According to the previous lemma, we have spectral representations

$$A = \lambda_1 \, |\boldsymbol{x}_1\rangle\langle\boldsymbol{x}_1| + \ldots + \lambda_n \, |\boldsymbol{x}_n\rangle\langle\boldsymbol{x}_n|$$

and

$$B = \mu_1 \, |\boldsymbol{x}_1\rangle\langle\boldsymbol{x}_1| + \ldots + \mu_n \, |\boldsymbol{x}_n\rangle\langle\boldsymbol{x}_n| \, .$$

Since the eigenvalues of U are of absolute value 1, it follows that the eigenvalues of A and B, i.e., numbers λ_i and μ_i, have absolute values of at most 1. But since A and B are self-adjoint, numbers λ_i and μ_i are also real. Thus, U can be written as

$$U = (\lambda_1 + i\mu_1) |\boldsymbol{x}_1\rangle\langle\boldsymbol{x}_1| + \ldots + (\lambda_n + i\mu_n) |\boldsymbol{x}_n\rangle\langle\boldsymbol{x}_n|,$$

where λ_j and μ_j are real numbers in the interval $[-1, 1]$. Because the eigenvalues of U have absolute value 1, we must have $\lambda_j^2 + \mu_j^2 = 1$. Thus, there exists a unique $\theta_j \in [0, 2\pi)$ for each j such that $\lambda_j = \cos\theta_j$ and $\mu_j = \sin\theta_j$. It follows that $\lambda_j + i\mu_j = \mathrm{e}^{i\theta_j}$ and U can be expressed as $U = \mathrm{e}^{iH}$, where

$$H = \theta_1 |\boldsymbol{x}_1\rangle\langle\boldsymbol{x}_1| + \ldots + \theta_n |\boldsymbol{x}_n\rangle\langle\boldsymbol{x}_n|.$$

Representation

$$U = \mathrm{e}^{iH} = \mathrm{e}^{i\theta_1} |\boldsymbol{x}_1\rangle\langle\boldsymbol{x}_1| + \ldots + \mathrm{e}^{i\theta_n} |\boldsymbol{x}_n\rangle\langle\boldsymbol{x}_n|$$

is called *the spectral representation* of unitary operator U. Operator H (or $-H$) is sometimes called the *Hamilton operator* which induces U. Notice also that the way we derived the spectral representation of a unitary operator gives us, as a byproduct, the knowledge that a unitary matrix has eigenvectors that form an orthonormal basis of H_n.

The spectral representation for a unitary operator could, of course, have been derived directly without using decomposition $U = A + iB$. Anyway, the spectral representation is useful when decomposing a unitary matrix into a product of simple unitary matrices. Notice also that if T_1 and T_2 commute, then clearly $\mathrm{e}^{i(T_1+T_2)} = \mathrm{e}^{iT_1}\mathrm{e}^{iT_2}$.

In the following examples, we will illustrate how the unitary matrices and the Hamiltonians that induce them are related.

Example 8.2.4. The *Hadamard-Walsh matrix*

$$W_2 = \frac{1}{\sqrt{2}}\begin{pmatrix} 1 & 1 \\ 1 & -1 \end{pmatrix}$$

is unitary but also self-adjoint. Thus, the decomposition $W_2 = A + iB$ is just W_2 itself. Matrix W_2 has two eigenvalues, -1 and 1, and the corresponding orthonormal eigenvectors can be chosen as $\boldsymbol{x}_{-1} = \frac{1}{\sqrt{4-2\sqrt{2}}}(1-\sqrt{2}, 1)^T$ and $\boldsymbol{x}_1 = \frac{1}{\sqrt{4+2\sqrt{2}}}(1+\sqrt{2}, 1)^T$. The corresponding projections can be expressed as

$$|\boldsymbol{x}_{-1}\rangle\langle\boldsymbol{x}_{-1}| = \begin{pmatrix} \frac{2-\sqrt{2}}{4} & -\frac{\sqrt{2}}{4} \\ -\frac{\sqrt{2}}{4} & \frac{2+\sqrt{2}}{4} \end{pmatrix}$$

and

$$|\boldsymbol{x}_1\rangle\langle\boldsymbol{x}_1| = \begin{pmatrix} \frac{2+\sqrt{2}}{4} & \frac{\sqrt{2}}{4} \\ \frac{\sqrt{2}}{4} & \frac{2-\sqrt{2}}{4} \end{pmatrix}.$$

The spectral representation of a *self-adjoint mapping* W_2 is, thus, given by

$$W_2 = 1 \cdot |\boldsymbol{x}_1\rangle\langle\boldsymbol{x}_1| + (-1) \cdot |\boldsymbol{x}_{-1}\rangle\langle\boldsymbol{x}_{-1}|,$$

which can be also written as $W_2 = e^{i \cdot 0} |\boldsymbol{x}_1\rangle\langle\boldsymbol{x}_1| + e^{i \cdot \pi} |\boldsymbol{x}_{-1}\rangle\langle\boldsymbol{x}_{-1}|$, so $W_2 = e^{iT}$, where

$$T = 0 \cdot |\boldsymbol{x}_1\rangle\langle\boldsymbol{x}_1| + \pi |\boldsymbol{x}_{-1}\rangle\langle\boldsymbol{x}_{-1}| = \pi |\boldsymbol{x}_{-1}\rangle\langle\boldsymbol{x}_{-1}|.$$

Example 8.2.5. Let θ be a real number and

$$R_\theta = \begin{pmatrix} \cos\theta & -\sin\theta \\ \sin\theta & \cos\theta \end{pmatrix}$$

be the *rotation matrix*. Clearly, R_θ is unitary. Now that we know that unitary matrices also have eigenvectors forming an orthonormal basis, we can find them directly without seeking a decomposition $R_\theta = A + iB$. It is an easy task to verify that the eigenvalues of R_θ are $e^{\pm i\theta}$, and the corresponding eigenvectors can be chosen as $\boldsymbol{x}_+ = \frac{1}{\sqrt{2}}(i, 1)$ and $\boldsymbol{x}_- = \frac{1}{\sqrt{2}}(-i, 1)^T$. The corresponding projections are given by

$$|\boldsymbol{x}_+\rangle\langle\boldsymbol{x}_+| = \begin{pmatrix} -\frac{1}{2} & \frac{i}{2} \\ \frac{i}{2} & \frac{1}{2} \end{pmatrix}$$

and

$$|\boldsymbol{x}_-\rangle\langle\boldsymbol{x}_-| = \begin{pmatrix} -\frac{1}{2} & -\frac{i}{2} \\ -\frac{i}{2} & \frac{1}{2} \end{pmatrix}.$$

Thus $R_\theta = e^{i\theta} |\boldsymbol{x}_+\rangle\langle\boldsymbol{x}_+| + e^{-i\theta} |\boldsymbol{x}_-\rangle\langle\boldsymbol{x}_-| = e^{iH_\theta}$, where

$$H_\theta = \begin{pmatrix} 0 & i\theta \\ -i\theta & 0 \end{pmatrix}.$$

Example 8.2.6. Let α and β be real numbers. The *phase shift matrix*

$$P_{\alpha,\beta} = \begin{pmatrix} e^{i\alpha} & 0 \\ 0 & e^{i\beta} \end{pmatrix}$$

is unitary, as is easily verified (Notice that these matrices are closely related to phase flips, see Example 2.1.4). The spectral decomposition is now trivial:

$$P_{\alpha,\beta} = e^{i\alpha} \begin{pmatrix} 1 & 0 \\ 0 & 0 \end{pmatrix} + e^{i\beta} \begin{pmatrix} 0 & 0 \\ 0 & 1 \end{pmatrix},$$

so $P_{\alpha,\beta} = e^{iH_{\alpha,\beta}}$, where

$$H_{\alpha,\beta} = \begin{pmatrix} \alpha & 0 \\ 0 & \beta \end{pmatrix}.$$

8.3 Quantum States as Hilbert Space Vectors

Let us return to the study of finite-level quantum systems. The mathematical description of such a system is based on an n-dimensional Hilbert space H_n. We choose an orthonormal basis $\{|\boldsymbol{x}_1\rangle, \ldots, |\boldsymbol{x}_n\rangle\}$ and call vectors $|\boldsymbol{x}_i\rangle$ *basis states.* A *state* of the system is a unit-length vector in H_n. States other than the basis ones are called *superpositions* of the basis states. Two states $|\boldsymbol{x}\rangle$ and $|\boldsymbol{y}\rangle$ are *equivalent* if $|\boldsymbol{x}\rangle = e^{i\theta}|\boldsymbol{y}\rangle$ for some real θ. Hereafter, we will regard equivalent states as equal.

A general state

$$\alpha_1 |\boldsymbol{x}_1\rangle + \ldots + \alpha_n |\boldsymbol{x}_n\rangle \tag{8.21}$$

determines a probability distribution: if the system in state (8.21) is observed, the system will be seen in a basis state $|\boldsymbol{x}_i\rangle$ with a probability of $|\alpha_i|^2$. The coefficients α_i are called *amplitudes.* We will study the observations in more detail in Section 8.3.2.

For now, we present a mathematical description of *compound quantum systems*: suppose we have n- and m-state distinguishable systems[9] with bases $\{|\boldsymbol{x}_1\rangle, \ldots, |\boldsymbol{x}_n\rangle\}$ of H_n and $\{|\boldsymbol{y}_1\rangle, \ldots, |\boldsymbol{y}_m\rangle\}$ of H_m respectively. The compound system which is made of these subsystems is described as tensor product $H_n \otimes H_m \simeq H_{nm}$. The basis states of $H_n \otimes H_m$ are

$$|\boldsymbol{x}_i\rangle \otimes |\boldsymbol{y}_j\rangle,$$

where $i \in \{1, \ldots, n\}$ and $j \in \{1, \ldots, m\}$. We also use notations $|\boldsymbol{x}_i\rangle \otimes |\boldsymbol{y}_j\rangle = |\boldsymbol{x}_i\rangle |\boldsymbol{y}_j\rangle = |\boldsymbol{x}_i, \boldsymbol{y}_j\rangle$. A general state $|\boldsymbol{z}\rangle$ of the compound system is a unit-length vector in H_{nm}. We say that a state $|\boldsymbol{z}\rangle$ is *decomposable,* if

$$|\boldsymbol{z}\rangle = |\boldsymbol{x}\rangle |\boldsymbol{y}\rangle$$

for some states $|\boldsymbol{x}\rangle \in H_n$ and $|\boldsymbol{y}\rangle \in H_m$. A state that is not decomposable is *entangled.* The inner product in space $H_n \otimes H_m$ is defined by

$$\langle \boldsymbol{x}_1 \otimes \boldsymbol{y}_1 \mid \boldsymbol{x}_2 \otimes \boldsymbol{y}_2 \rangle = \langle \boldsymbol{x}_1 \mid \boldsymbol{x}_2 \rangle \langle \boldsymbol{y}_1 \mid \boldsymbol{y}_2 \rangle.$$

Example 8.3.1. A two-level quantum system is called a *qubit.* The orthonormal basis vectors chosen for H_2 are usually denoted by $|0\rangle$ and $|1\rangle$ and are identified with logical 0 and 1. A compound system of m qubits is called a *quantum register of length* m and is described by a Hilbert space having dimension 2^m and basis

$$\{|x_1\rangle \ldots |x_m\rangle \mid x_i \in \{0,1\}\}.$$

We can identify the binary sequence $x_1, \ldots, x_m$ and number $x_1 2^{m-1} + x_2 2^{m-2} + \ldots + x_{m-1} 2 + x_m$. Using this identification, the basis of H_{2^m} can be expressed as

$$\{|0\rangle, |1\rangle, \ldots, |2^m - 1\rangle\}.$$

[9] The requirement that the systems are distinguishable is essential here. A compound system of two *identical* subsystems has a different description.

8.3.1 Quantum Time Evolution

Our next task is to describe how quantum systems change in time. For that purpose, we will assume, as is usually done, that there is some function $U_t : H_n \to H_n$ which depends on the time t, and describes the time evolution of the system.[10] In other words, by denoting the state of the system at time t by $\boldsymbol{x}(t)$,

$$\boldsymbol{x}(t) = U_t \boldsymbol{x}(0).$$

We will now list some requirements that are usually put on the *time-evolution mapping* U_t. The very first stress put on U_t is that it should map unit-length states to unit-length states; that is, it should preserve the norm:

1. For each $t \in \mathbb{R}$ and each $\boldsymbol{x} \in H_n$, $||U_t \boldsymbol{x}|| = ||\boldsymbol{x}||$.

Notice that the above requirement is quite mathematical; it just states that, if the coefficients of $\boldsymbol{x}$ determine a probability distribution as in (8.21), then the coefficients of $U\boldsymbol{x}$ should do so as well. The second requirement, on the other hand, is that the superposition (8.21) should strongly resemble the probability distribution, so strongly that each U_t operates on (8.21) in such a way that each basis state evolves independently:

2. For each t, $U_t(\alpha_1 |\boldsymbol{x}_1\rangle + \ldots + \alpha_n |\boldsymbol{x}_n\rangle) = \alpha_1 U_t |\boldsymbol{x}_1\rangle + \ldots + \alpha_n U_t |\boldsymbol{x}_n\rangle$. This means that each U_t is linear.

We could seriously argue about the above requirement,[11] but that is not the purpose of the present book. The next requirement is that U_t can always be decomposed:

3. For all t_1 and $t_2 \in \mathbb{R}$, $U_{t_1+t_2} = U_{t_1} U_{t_2}$.

Finally, we require that the time evolution must be smooth:

4. For each $t_0 \in \mathbb{R}$, $\lim_{t \to t_0} U_t \boldsymbol{x}(0) = \lim_{t \to t_0} \boldsymbol{x}(t) = \boldsymbol{x}(t_0)$.

From the above requirements, we can derive the following characteristic result.

Lemma 8.3.1. *A time evolution mapping U_t satisfying 1–3 is a unitary operator.*

Proof. Condition 2 states that U_t is an operator, and from condition 1, it follows directly that each U_t is injective. Requirement 3 implies that $U_0 = U_0^2$. But $U_0 |\boldsymbol{x}\rangle = U_0^2 |\boldsymbol{x}\rangle$ implies, by the injectivity, that $|\boldsymbol{x}\rangle = U_0 |\boldsymbol{x}\rangle$ for any $|\boldsymbol{x}\rangle$, which is to say that U_0 is the identity mapping. By again using requirement 3, we see that $|\boldsymbol{x}\rangle = U_t U_{-t} |\boldsymbol{x}\rangle$, i.e., each U_t is also surjective. According to requirement 1, each U_t preserves norm, and by Lemma 8.2.7 it follows that each u_t is unitary. □

[10] This assumption is referred to as the *causality principle*.

[11] If a state (8.21) is not just a generalization of a probability distribution, why should it evolve like one?

Condition 4 is still unused. Its place is found after the following theorem by M. Stone; see [65].

Theorem 8.3.1. *If each U_t satisfies 1–4, then there is a unique self-adjoint operator H such that*

$$U_t = e^{-itH}.$$

Recall that the exponential function of a self-adjoint operator A can be defined by using the spectral representation (8.19), or even as series

$$e^A = I + A + \frac{1}{2!}A^2 + \frac{1}{3!}A^3 + \ldots.$$

It is clear that the definitions coincide in a finite-dimensional H_n. By Stone's theorem, the time evolution can be expressed as

$$\boldsymbol{x}(t) = e^{-itH}\boldsymbol{x}(0),$$

from which we get

$$\frac{d}{dt}\boldsymbol{x}(t) = -iHe^{-itH}\boldsymbol{x}(0) = -iH\boldsymbol{x}(t)$$

by component-wise differentiation. This can also be written as

$$i\frac{d}{dt}\boldsymbol{x}(t) = H\boldsymbol{x}(t). \tag{8.22}$$

Differential equation (8.22) is known as the *abstract Schrödinger equation*, and the operator H is called the *Hamilton operator* of the quantum system. In a typical situation, the Hamilton operator (which represents the *total energy* of the system) is known, and the Schrödinger equation is used to determine the *energy eigenstates*, which form an example of the basis states of a quantum system.

Remark 8.3.1. The time evolution which is described by operators U_t is continuous, but, in connection to computational aspects, we are interested in the system state at discrete time points $t_1, t_2, t_3, \ldots$. Therefore, we will regard the quantum system evolution as a sequence of unit-length vectors $\boldsymbol{x}$, $U_1\boldsymbol{x}$, $U_2U_1\boldsymbol{x}$, $U_3U_2U_1\boldsymbol{x}$, ..., where each U_i is unitary.

8.3.2 Observables

We again study a general state

$$|\boldsymbol{x}\rangle = \alpha_1 |\boldsymbol{x}_1\rangle + \ldots + \alpha_n |\boldsymbol{x}_n\rangle. \tag{8.23}$$

The basis states were introduced in connection with the distinguishable properties which we are interested in, and a general state, i.e., a superposition of basis states, induces a probability distribution of the basis states: a basis state $\boldsymbol{x}_i$ is observed with a probability of $|\alpha_i|^2$. This will be generalized as follows:

Definition 8.3.1. *An observable $E = \{E_1, \ldots E_m\}$ is a collection of mutually orthogonal subspaces of H_n such that*

$$H_n = E_1 \oplus \ldots \oplus E_m.$$

In the above definition, inequality $m \leq n$ must, of course, hold. We equip the subspaces E_i with distinct real number "labels" $\theta_1, \ldots, \theta_m$. For each vector $\boldsymbol{x} \in H_n$, there is a unique representation

$$\boldsymbol{x} = \boldsymbol{x}_1 + \ldots + \boldsymbol{x}_m \tag{8.24}$$

such that $\boldsymbol{x}_i \in E_i$. Instead of observing spaces E_i, we can talk about observing the labels θ_i: we say that by observing E, value θ_i will be seen with a probability of $||\boldsymbol{x}_i||^2$.[12]

Example 8.3.2. The notion of observing the basis states $\boldsymbol{x}_i$ is a special case: the observable E can be defined as

$$E = \{E_1, \ldots, E_n\},$$

where each E_i is the one-dimensional subspace spanned by $\boldsymbol{x}_i$. We can, for instance, equip E_i with the label i. If the system is in state (8.23), the value i is observed with a probability of

$$||\alpha_i \boldsymbol{x}_i||^2 = |\alpha_i|^2.$$

Another view of the observables can be achieved in the following way: if $H_n = E_1 \oplus \ldots \oplus E_m$, let P_{E_i} be the projection on the subspaces E_i. Thus, we may also think of observables as (sometimes partially defined) mappings $E : \mathbb{R} \to L(H)$ that associate a projection $E(\theta_i) = P_{E_i}$ to each label θ_i. Following the notation (8.24),

$$P_{E_i}(\boldsymbol{x}) = \boldsymbol{x}_i,$$

and the probability of observing the label θ_i when the system state is $|\boldsymbol{x}\rangle$, is given by

$$||\boldsymbol{x}_i||^2 = \langle P_{E_i}\boldsymbol{x} \mid P_{E_i}\boldsymbol{x}\rangle = \langle \boldsymbol{x} \mid P_{E_i}^2 \boldsymbol{x}\rangle = \langle \boldsymbol{x} \mid P_{E_i}\boldsymbol{x}\rangle = \langle \boldsymbol{x} \mid E(\theta_i)\boldsymbol{x}\rangle.$$

Mapping $E : \mathbb{R} \to L(H)$ has other interesting properties: the probability of observing either θ_i or θ_j is clearly

$$\langle \boldsymbol{x} \mid E(\theta_i)\boldsymbol{x}\rangle + \langle \boldsymbol{x} \mid E(\theta_j)\boldsymbol{x}\rangle = \langle \boldsymbol{x} \mid (E(\theta_i) + E(\theta_j))\boldsymbol{x}\rangle,$$

[12] The original starting point is, of course, controversial: instead of talking about observing subspaces, one talks about measuring some physical quantity and obtaining a real number as the outcome. However, here it seems to be logically more consistent to introduce these quantity values as labels of subspaces.

and the probability that *some* θ_i is observed is

$$1 = \langle \boldsymbol{x} \mid I\boldsymbol{x} \rangle.$$

Thus, we may extend E to be defined on *sets* of real numbers by setting $E(\{\theta_i, \theta_j\}) = E(\theta_i) + E(\theta_j)$ if $\theta_i \neq \theta_j$. We can easily see that, as a mapping from subsets of $\mathbb{R}$ to $L(H)$, E satisfies the following conditions:

1. For each $X \subset \mathbb{R}$, $E(X)$ is a projection.
2. $E(\mathbb{R}) = I$.
3. $E(\cup X_i) = \sum E(X_i)$ for a disjoint sequence X_i of sets.

A mapping E that satisfies the above conditions is called a *projection-valued measure.*[13] Viewing an observable as a projection-valued measure represents an alternative way of thinking than Definition (8.3.1). In that definition, we merely equipped each subspace with a real number, a label, which is thought to be observed instead of the actual subspace. Here we associate a projection which defines a subspace to a set of labels (real numbers). If X is a set of real numbers, then $\langle \boldsymbol{x} \mid E(X)\boldsymbol{x} \rangle$ is just the probability of seeing a number in the X.

It follows from Condition 3 that, if X and Y are disjoint sets, then the corresponding projections $E(X)$ and $E(Y)$ project onto mutually orthogonal subspaces. In fact, since $E(X)(H_n)$ is a subspace of $E(X \cup Y)(H_n)$, we must have $E(X)E(X \cup Y) = E(X)$ and; therefore, $E(X) = E(X)E(X \cup Y) = E(X) + E(X)E(Y)$, so $E(X)E(Y) = 0$.

The third and perhaps the most traditional view of the observables can be achieved by regarding an observable as a self-adjoint operator

$$A = \theta_1 E(\theta_1) + \ldots + \theta_m E(\theta_m),$$

where $E(\theta_i) = E(\{\theta_i\})$ are projections to mutually orthogonal subspaces. Recalling that the probability of observing θ_i when the system is in state $\boldsymbol{x}$, is $\langle \boldsymbol{x} \mid E(\theta_i)\boldsymbol{x} \rangle$, we can compute the *expected value* of an observable A in state $\boldsymbol{x}$. The expected value is:

$$\begin{aligned} \mathbb{E}_{\boldsymbol{x}}(A) &= \theta_1 \langle \boldsymbol{x} \mid E(\theta_1)\boldsymbol{x} \rangle + \ldots + \theta_m \langle \boldsymbol{x} \mid E(\theta_m)\boldsymbol{x} \rangle \\ &= \langle \boldsymbol{x} \mid (\theta_1 E(\theta_1) + \ldots + \theta_m E(\theta_m))\boldsymbol{x} \rangle = \langle \boldsymbol{x} \mid A\boldsymbol{x} \rangle. \end{aligned}$$

It now seems to be a good moment to collect together all the viewpoints of observables which we have so far.

- An observable can be defined as a collection of mutually orthogonal subspaces $\{E_1, \ldots, E_m\}$ such that $H = E_1 \oplus \ldots \oplus E_m$. Each subspace is

[13] The notion of a projection-valued measure is not, in general, defined on all subsets of $\mathbb{R}$, but on a σ-algebra which has enough "regular features". Typically, the *Borel* subsets of $\mathbb{R}$ will suffice. However, now we are interested in finite-level quantum systems, so there are only finitely many real numbers associated with each observable.

equipped with a real number label θ_i. This point of view most closely resembles the original idea of thinking about a quantum state as a generalization of a probability distribution. In fact, observing the state of a quantum system can be seen as learning the value of an observable, which is defined as a collection of one-dimensional subspaces spanned by the basis states.
- An observable can be seen as a projection-valued measure E which maps a set of labels to a projection that defines a subspace. This viewpoint offers us a method for generalizing the notion of an observable into *positive operator-valued* measure.
- The traditional view of an observable is to define an observable as a self-adjoint operator by the spectral representation

$$A = \theta_1 E(\theta_1) + \ldots + \theta_m E(\theta_m).$$

All these viewpoints are logically quite equal. In what follows, we will use all of them, choosing the one which is best-suited for each situation.

8.3.3 The Uncertainty Principles

In this section, we first present an interesting classical result connected to the simultaneous measurements of two observables. We will view the observables as self-adjoint mappings. We say that two observables A and B are *commutative*, if $AB = BA$. We also define the *commutator* of A and B as a mapping $[A, B] = AB - BA$. Recall that, for a given state $\boldsymbol{x}$, the expected value of observable A is given by $\mathbb{E}_{\boldsymbol{x}}(A) = \langle \boldsymbol{x} \mid A\boldsymbol{x} \rangle$. For short, we denote $\mu = \mathbb{E}_{\boldsymbol{x}}(A)$. The *variance* of A (in state $\boldsymbol{x}$) is the expected value of observable $(A - \mu)^2$. Notice that, if A is self-adjoint, then $(A - \mu)^2$ is also self-adjoint, and, thus, it also defines an observable. With these notations, the variance can be expressed as

$$\begin{aligned} \mathrm{Var}_{\boldsymbol{x}}(A) &= \mathbb{E}_{\boldsymbol{x}}((A-\mu)^2) = \langle \boldsymbol{x} \mid (A-\mu)^2 \boldsymbol{x} \rangle \\ &= \langle (A-\mu)\boldsymbol{x} \mid (A-\mu)\boldsymbol{x} \rangle = \|(A-\mu)\boldsymbol{x}\|^2. \end{aligned} \tag{8.25}$$

Theorem 8.3.2 (The Uncertainty Principle). *For any observables A, B and any state $\boldsymbol{x}$,*

$$\mathrm{Var}_{\boldsymbol{x}}(A)\mathrm{Var}_{\boldsymbol{x}}(B) \geq \frac{1}{4} \left| \langle \boldsymbol{x} \mid [A, B]\boldsymbol{x} \rangle \right|^2.$$

Proof. Notice first that, for self-adjoint A and B, $[A, B]^* = -[A, B]$, so the commutator of A and B can be written as $[A, B] = iC$, where $C = -i[A, B]$ is self-adjoint. If μ_A and μ_B are the expected values of A and B respectively, we get

$$\begin{aligned} \mathrm{Var}_{\boldsymbol{x}}(A)\mathrm{Var}_{\boldsymbol{x}}(B) &= \|(A-\mu_A)\boldsymbol{x}\|^2 \|(B-\mu_B)\boldsymbol{x}\|^2 \\ &\geq \left| \langle (A-\mu_A)\boldsymbol{x} \mid (B-\mu_B)\boldsymbol{x} \rangle \right|^2 \end{aligned}$$

using representation (8.25) and the Cauchy-Schwartz inequality. To shorten the notations, we write $A_1 = A - \mu_A$ and $B_1 = B - \mu_B$. Clearly, A_1 and B_1 are also self-adjoint and it is a straightforward task to verify that $[A_1, B_1] = [A, B]$ and that $A_1B_1 + B_1A_1$ is self-adjoint. Thus,

$$
\begin{aligned}
&|\langle A_1\boldsymbol{x} \mid B_1\boldsymbol{x}\rangle|^2 = |\langle \boldsymbol{x} \mid A_1B_1\boldsymbol{x}\rangle|^2 \\
&= \left|\langle \boldsymbol{x} \mid \tfrac{1}{2}(A_1B_1 + B_1A_1)\boldsymbol{x} + \tfrac{1}{2}(A_1B_1 - B_1A_1)\boldsymbol{x}\rangle\right|^2 \\
&= \frac{1}{4}\,|\langle \boldsymbol{x} \mid (A_1B_1 + B_1A_1)\boldsymbol{x}\rangle + \langle \boldsymbol{x} \mid [A, B]\boldsymbol{x}\rangle|^2 \\
&= \frac{1}{4}\,|\langle \boldsymbol{x} \mid (A_1B_1 + B_1A_1)\boldsymbol{x}\rangle + i\langle \boldsymbol{x} \mid C\boldsymbol{x}\rangle|^2,
\end{aligned}
$$

where $C = -i[A, B]$ is a self-adjoint operator. Because of the self-adjointness of the above operators, both inner products are real. Therefore,

$$
\begin{aligned}
\mathrm{Var}_{\boldsymbol{x}}(A)\mathrm{Var}_{\boldsymbol{x}}(B) &\geq \frac{1}{4}\,|\langle \boldsymbol{x} \mid (A_1B_1 + B_1A_1)\boldsymbol{x}\rangle + i\langle \boldsymbol{x} \mid C\boldsymbol{x}\rangle|^2 \\
&= \frac{1}{4}(\langle \boldsymbol{x} \mid (A_1B_1 + B_1A_1)\boldsymbol{x}\rangle^2 + \langle \boldsymbol{x} \mid C\boldsymbol{x}\rangle^2) \\
&\geq \frac{1}{4}\langle \boldsymbol{x} \mid C\boldsymbol{x}\rangle^2,
\end{aligned}
$$

which can also be written as

$$
\mathrm{Var}_{\boldsymbol{x}}(A)\mathrm{Var}_{\boldsymbol{x}}(B) \geq \frac{1}{4}\,|\langle \boldsymbol{x} \mid [A, B]\boldsymbol{x}\rangle|^2,
$$

as claimed. □

Remark 8.3.2. A classical example of noncommutative observables are position and momentum. It turns out that the commutator of these two observables is a homothety, i.e., multiplication by a nonzero constant, so Lemma 8.3.2 demonstrates that, in any state, the product of the variances of position and momentum has a positive lower bound. This is known as *Heisenberg's uncertainty principle.*

The uncertainty principle (Theorem 8.3.2) was criticized by D. Deutsch [30] on the grounds that the lower bound provided by Theorem 8.3.2 is not generally fixed, but depends on the state $\boldsymbol{x}$. Deutsch himself in [30] gave a state-independent lower bound for the sum of the *uncertainties* of two observables. Here we present an improved version of this bound courtesy of Maassen and Uffink [56].

We say that a sequence $(p_1, \ldots, p_n)$ of real numbers is a *probability distribution* if $p_i \geq 0$ and $p_1 + \ldots + p_n = 1$. For a probability distribution $P = (p_1, \ldots, p_n)$, the *Shannon entropy* of distribution is defined to be

$$
H(P) = -(p_1 \log p_1 + \ldots + p_n \log p_n),
$$

where $0 \cdot \log 0$ is defined to be 0. For basic properties of Shannon entropy, see Section 9.5.

Let A and B be some observables of an n-state quantum system (again regarded as self-adjoint operators). As discussed before, there are orthonormal bases of H_n consisting of the eigenvectors of A and B. By denoting these bases by $\{\boldsymbol{a}_1, \ldots, \boldsymbol{a}_n\}$ and $\{\boldsymbol{b}_1, \ldots, \boldsymbol{b}_n\}$, we get spectral representations of A and B:

$$A = \lambda_1 \, |\boldsymbol{a}_1\rangle\langle\boldsymbol{a}_1| + \ldots + \lambda_n \, |\boldsymbol{a}_n\rangle\langle\boldsymbol{a}_n|,$$
$$B = \mu_1 \, |\boldsymbol{b}_1\rangle\langle\boldsymbol{b}_1| + \ldots + \mu_n \, |\boldsymbol{b}_n\rangle\langle\boldsymbol{b}_n|,$$

where λ_i and μ_i are the eigenvalues of A and B, respectively.

With respect to a fixed state $|\boldsymbol{x}\rangle \in H_n$, observables A and B define probability distributions $P = (p_1, \ldots, p_n)$ and $Q = (q_1, \ldots, q_n)$ that describe the probabilities of seeing labels $\lambda_1, \ldots, \lambda_n$ and $\mu_1, \ldots, \mu_n$. Notice that the projections $E(\lambda_i)$ and $E(\mu_i)$ are $|\boldsymbol{a}_i\rangle\langle\boldsymbol{a}_i|$ and $|\boldsymbol{b}_i\rangle\langle\boldsymbol{b}_i|$ respectively. Therefore, by measuring observable A when the system is in state $|\boldsymbol{x}\rangle$, label λ_i can be observed with a probability of

$$p_i = \langle\boldsymbol{x}\,||\boldsymbol{a}_i\rangle\langle\boldsymbol{a}_i|\,\boldsymbol{x}\rangle = \langle\boldsymbol{x} \mid \langle\boldsymbol{a}_i \mid \boldsymbol{x}\rangle\boldsymbol{a}_i\rangle = |\langle\boldsymbol{a}_i \mid \boldsymbol{x}\rangle|^2,$$

and by measuring B, label μ_i is seen with a probability of $q_i = |\langle\boldsymbol{b}_i \mid \boldsymbol{x}\rangle|^2$.

The probability distributions $P = (p_1, \ldots, p_m)$ and $Q = (q_1, \ldots, q_m)$ obey the following state-independent law:

Theorem 8.3.3 (Entropic Uncertainty Relation).

$$H(P) + H(Q) \geq -2 \log c,$$

where $c = \max_{i,j} |\langle\boldsymbol{a}_i \mid \boldsymbol{b}_j\rangle|$.

Proof. The expressions $H(P)$ and $H(Q)$ seem somewhat difficult to estimate, so we should find something to replace them. Let $r > 0$ and study the expression

$$\log\Big(\sum_{i=1}^{n} p_i^{r+1}\Big)^{\frac{1}{r}} = \frac{1}{r}\log\Big(\sum_{i=1}^{n} p_i^{r+1}\Big). \tag{8.26}$$

We claim that

$$\lim_{r\to 0}\frac{1}{r}\log\Big(\sum_{i=1}^{n} p_i^{r+1}\Big) = \sum_{i=1}^{n} p_i \log p_i. \tag{8.27}$$

In fact, since $\sum_{i=1}^{n} p_i = 1$, the nominator and numerator of the left-hand side of (8.27) tend to 0 as r does. Therefore, L'Hospital rule applies and

$$\lim_{r\to 0}\frac{\log(\sum_{i=1}^{n} p_i^{r+1})}{r} = \lim_{r\to 0}\frac{1}{\sum_{i=1}^{n} p_i^{r+1}}\sum_{i=1}^{n} p_i^{r+1}\log p_i = \sum_{i=1}^{n} p_i \log p_i.$$

Notice also that (8.27) implies that

$$\lim_{r\to 0}\left(\sum_{i=1}^{n} p_i^{r+1}\right)^{\frac{1}{r}} = \prod_{i=1}^{n} p_i^{p_i},$$

and that the claim is equivalent to $\exp(-H(P) - H(Q)) \le c^2$, which can be written as

$$\prod_{i=1}^{n} p_i^{p_i} \prod_{i=1}^{n} q_i^{q_i} \le c^2. \tag{8.28}$$

Our strategy is to replace the products in (8.28) with expressions similar to those ones in (8.26), estimate them, and obtain (8.28) letting $r \to 0$.

To estimate (8.26), we use a theorem by M. Riesz [73]:[14] If $T : H_n \to H_n$ is a unitary matrix, $\boldsymbol{y} = T\boldsymbol{x}$, and $1 \le a \le 2$, then

$$\left(\sum_{i=1}^{n} |y_k|^{\frac{a}{a-1}}\right)^{\frac{a-1}{a}} \le c^{\frac{2-a}{a}} \left(\sum_{i=1}^{n} |x_i|^a\right)^{\frac{1}{a}},$$

where $c = \max |T_{ij}|$.

By choosing $x_i = \langle \boldsymbol{a}_i \mid \boldsymbol{x}\rangle$, $T_{ij} = \langle \boldsymbol{b}_i \mid \boldsymbol{a}_j\rangle$, we have $y_i = \langle \boldsymbol{b}_i \mid \boldsymbol{x}\rangle$. Clearly, T is unitary. We can now write the theorem in form

$$\left(\sum_{i=1}^{n} |\langle \boldsymbol{b}_i \mid \boldsymbol{x}\rangle|^{\frac{a}{a-1}}\right)^{\frac{a-1}{a}} \left(\sum_{i=1}^{n} |\langle \boldsymbol{a}_i \mid \boldsymbol{x}\rangle|^{a}\right)^{-\frac{1}{a}} \le c^{\frac{2-a}{a}}. \tag{8.29}$$

Since (8.29) can also be raised to some positive power k, we will try to fix values such that (8.29) would begin to resemble the right-hand side of (8.26). Therefore, we will search for numbers r, s, and k such that $\frac{a}{a-1} = 2(s+1)$, $k\frac{a-1}{a} = \frac{1}{s}$, $a = 2(r+1)$, $-\frac{k}{a} = \frac{1}{r}$, and $k\frac{2-a}{a} = 2$. To satisfy $1 \le a \le 2$, we must choose $r \in [-\frac{1}{2}, 0]$. A simple calculation shows that choosing $s = -\frac{r}{2r+1}$, $k = -\frac{2r+2}{r}$ will suffice. Thus, (8.29) can be rewritten as

$$\left(\sum_{i=1}^{n} q_i^{s+1}\right)^{\frac{1}{s}} \left(\sum_{i=1}^{n} p_i^{r+1}\right)^{\frac{1}{r}} \le c^2,$$

and, by letting $r \to 0$, we obtain the desired estimation. □

Remark 8.3.3. The lower bound of Theorem 8.3.3 does not depend on the system state $\boldsymbol{x}$, but it may depend on the representation of observables

$$A = \lambda_1 \,|\boldsymbol{a}_1\rangle\langle \boldsymbol{a}_1| + \ldots + \lambda_n \,|\boldsymbol{a}_n\rangle\langle \boldsymbol{a}_n|$$

and

[14] Riesz's theorem is a special case of Stein's interpolation formula. The elementary proof given by Riesz [73] is based on Hölder's inequality.

$$B = \mu_1 \, |\boldsymbol{b}_1\rangle\langle\boldsymbol{b}_1| + \ldots + \mu_n \, |\boldsymbol{b}_n\rangle\langle\boldsymbol{b}_n| \, .$$

The independence of the representation is obtained if both observables are non-degenerate. Then, all the vectors $\boldsymbol{a}_i$ and $\boldsymbol{b}_i$ are determined uniquely up to a multiplicative factor having absolute values of 1; hence $c = \max |\langle \boldsymbol{b}_i \mid \boldsymbol{a}_j \rangle|$ is also uniquely determined. Notice also that the degeneration of an observable means that there are less than n values to be observed, so naturally the entropy decreases.

Example 8.3.3. Consider a system consisting of m qubits. One orthonormal basis of H_{2^m} is the familiar computational basis $B_1 = \{|x_1 \ldots x_m\rangle \mid x_i \in \{0,1\}\}$, and another one is $B_2 = \{H\,|\boldsymbol{x}\rangle \mid |\boldsymbol{x}\rangle \in B_1\}$ which is obtained by applying Hadamard-Walsh transform (see Section 4.1.2 or 9.2.3)

$$H\,|\boldsymbol{x}\rangle = \frac{1}{\sqrt{2^m}} \sum_{\boldsymbol{y} \in \mathbb{F}_2^m} (-1)^{\boldsymbol{x}\cdot\boldsymbol{y}} \, |\boldsymbol{y}\rangle \, .$$

on B_1. We will study observables A_1 and A_2 which are determined by B_1 and B_2,

$$A_i = \sum_{\boldsymbol{x} \in B_i} c_{\boldsymbol{x}}^{(i)} \, |\boldsymbol{x}\rangle\langle\boldsymbol{x}| \, .$$

Now we do not care very much about the labels $c_{\boldsymbol{x}}^{(i)}$; they can be fixed in any manner such that both sequences $c_{\boldsymbol{x}}^{(i)}$ consist of 2^m distinct real numbers. This, in fact, is to regard an observable as a decomposition of H_{2^m} into 2^m mutually orthogonal one-dimensional subspaces. For any $|\boldsymbol{x}\rangle \in B_1$ and $|\boldsymbol{x}'\rangle \in B_2$, we have

$$|\langle \boldsymbol{x} \mid \boldsymbol{x}' \rangle| = \frac{1}{\sqrt{2^m}},$$

as is easily verified. Thus, the entropic uncertainty relation gives

$$H(A_1) + H(A_2) \geq m.$$

That is, the sum of the uncertainties of observing an m-qubit state in bases B_1 and B_2 mutually is at least m bits! On the other hand, the uncertainty of a single observation cannot naturally be more than m bits: an observation with respect to any basis will give us a sequence $\boldsymbol{x} = x_1 \ldots x_m \in \{0,1\}^m$ with some probabilities of $p_{\boldsymbol{x}}$ such that $\sum_{\boldsymbol{x} \in \mathbb{F}_2^m} p_{\boldsymbol{x}} = 1$. The Shannon entropy is given by

$$-\sum_{\boldsymbol{x} \in \mathbb{F}_2^m} p_{\boldsymbol{x}} \log p_{\boldsymbol{x}},$$

which achieves the maximum at $p_{\boldsymbol{x}} = \frac{1}{2^m}$ for each $\boldsymbol{x} \in \{0,1\}^m$. Thus, the maximum uncertainty is m bits.

The lower bound of the uncertainty principle (Theorem 8.3.2) depends on the commutator $[A, B]$ and on the state $|\boldsymbol{x}\rangle$. If $[A, B] = 0$, then the lower is always trivial. Does a similar result hold also for the entropic uncertainty relation? The answer is yes: if A and B commute, then, according to Lemma 8.2.9, A and B share an orthonormal eigenvector basis. But then $c = \max |\langle \boldsymbol{b}_i \mid \boldsymbol{a}_j \rangle| = 1$ and the lower bound is trivial.

8.4 Quantum States as Operators

Let us study a system of two qubits in state

$$\frac{1}{\sqrt{2}}(|0\rangle\,|0\rangle + |1\rangle\,|1\rangle). \tag{8.30}$$

In the introductory section we learned that this state is entangled; there is no way to write (8.30) as a product of two single-qubit states. This also means that the representation of quantum systems as Hilbert state vectors does not offer us any vocabulary to speak about the state of the first qubit if the compound system is in state (8.30). In this section, we will generalize the notion of a quantum state.

8.4.1 Density Matrices

Let $\{\boldsymbol{b}_1, \ldots, \boldsymbol{b}_n\}$ be a fixed basis of H_n. In this section, the coordinate representation of a vector $\boldsymbol{x} = x_1\boldsymbol{b}_1 + \ldots + x_n\boldsymbol{b}_n$ is interpreted as a column vector $(x_1, \ldots, x_n)^T$, and we define the *dual vector* of $|\boldsymbol{x}\rangle$ as a row vector $\langle \boldsymbol{x}| = (x_1^*, \ldots, x_n^*)$. Notice that, for any $\boldsymbol{x} = x_1\boldsymbol{b}_1 + \ldots x_n\boldsymbol{b}_n$ and $\boldsymbol{y} = y_1\boldsymbol{b}_1 + \ldots + y_n\boldsymbol{b}_n$, $\langle \boldsymbol{x}|\,|\boldsymbol{y}\rangle$ as a matrix product looks like

$$\langle \boldsymbol{x}|\,|\boldsymbol{y}\rangle = (x_1^*, \ldots, x_n^*)\begin{pmatrix} y_1 \\ \vdots \\ y_n \end{pmatrix} = x_1^* y_1 + \ldots + x_n^* y_n = \langle \boldsymbol{x} \mid \boldsymbol{y} \rangle.$$

Inspired by the word "bracket", we will call the row vectors $\langle \boldsymbol{x}|$ *bra*-vectors and column vectors $|\boldsymbol{y}\rangle$ *ket*-vectors.[15]

Remark 8.4.1. A coordinate independent notion of dual vectors can be achieved by using the theorem of Fréchet and F. Riesz. Any vector $\boldsymbol{x} \in H_n$ defines a linear function $f_{\boldsymbol{x}} : H_n \to \mathbb{C}$ by

$$f_{\boldsymbol{x}}(\boldsymbol{y}) = \langle \boldsymbol{x} \mid \boldsymbol{y} \rangle. \tag{8.31}$$

A linear function $f : H_n \to \mathbb{C}$ is called a *functional*. Functionals form a vector space H_n^*, a so-called *dual space* of H_n, where addition and scalar

[15] This terminology was first used by P. A. M. Dirac.

multiplication are defined pointwise. The Fréchet-Riesz's theorem states that, for any functional $f : H_n \to \mathbb{C}$, there is a unique vector $\boldsymbol{x} \in H_n$ such that $f = f_{\boldsymbol{x}}$. According to this point of view, the dual vector of $\boldsymbol{x}$ is the functional $f_{\boldsymbol{x}}$ that satisfies (8.31).

For any state (vector of unit length) $\boldsymbol{x} = x_1\boldsymbol{b}_1 + \ldots + x_n\boldsymbol{b}_n \in H_n$, we define the *density matrix* belonging to $\boldsymbol{x}$ by

$$
|\boldsymbol{x}\rangle \otimes \langle\boldsymbol{x}| = \begin{pmatrix} x_1 \\ x_2 \\ \vdots \\ x_n \end{pmatrix} \otimes (x_1^*, x_2^*, \ldots, x_n^*)
$$

$$
= \begin{pmatrix} |x_1|^2 & x_1x_2^* & \ldots & x_1x_n^* \\ x_2x_1^* & |x_2|^2 & \ldots & x_2x_n^* \\ \vdots & \vdots & \ddots & \vdots \\ x_nx_1^* & x_nx_2^* & \ldots & |x_n|^2 \end{pmatrix}.
$$

Remark 8.4.2. For density matrices, there is also a coordinate-independent viewpoint: it is not difficult to verify that $|\boldsymbol{x}\rangle \otimes \langle\boldsymbol{x}|$ is a matrix representing a projection onto the one-dimensional subspace generated by $\boldsymbol{x}$, and we have already used notation $|\,\boldsymbol{x}\rangle\langle\boldsymbol{x}\,|$ to stand for such a projection. To shorten the notations, we will omit $\otimes$ and write $|\boldsymbol{x}\rangle \otimes \langle\boldsymbol{x}| = |\,\boldsymbol{x}\rangle\langle\boldsymbol{x}\,|$ hereafter. Recall also that we called states $|\boldsymbol{x}\rangle$ and $|\boldsymbol{x}_1\rangle$ equal, if $\boldsymbol{x} = e^{i\theta}\boldsymbol{x}_1$ for some real number θ, but the density matrices are robust against factor $e^{i\theta}$, i.e., $|\,\boldsymbol{x}_1\rangle\langle\boldsymbol{x}_1\,| = |\,\boldsymbol{x}\rangle\langle\boldsymbol{x}\,|$. On the other hand, for a one-dimensional subspace W, we can pick a unit-length vector $\boldsymbol{x}$ which generates W. This vector is unique up to the multiplicative constant $e^{i\theta}$, and the projection onto W is $|\,\boldsymbol{x}\rangle\langle\boldsymbol{x}\,|$. Thus, there is a bijective correspondence between the states and the density matrices belonging to unit vectors $\boldsymbol{x}$.

We list three important properties of density matrices $|\,\boldsymbol{x}\rangle\langle\boldsymbol{x}\,|$ as follows:

- Since $||\boldsymbol{x}|| = 1$, matrices $|\,\boldsymbol{x}\rangle\langle\boldsymbol{x}\,|$ have unit trace (recall that the trace is the sum of the diagonal elements).
- As projections, matrices $|\,\boldsymbol{x}\rangle\langle\boldsymbol{x}\,|$ are self-adjoint.
- Matrices $|\,\boldsymbol{x}\rangle\langle\boldsymbol{x}\,|$ are positive, since $(|\,\boldsymbol{x}\rangle\langle\boldsymbol{x}\,|)^2 = |\,\boldsymbol{x}\rangle\langle\boldsymbol{x}\,|$.

Using these properties, we generalize the notion of a state:

Definition 8.4.1. *A state of a quantum system is a positive, unit-trace self-adjoint operator in H_n. Such an operator is called a density operator.*

Hereafter, any matrix representing a state is called a density matrix as well. Spectral representation offers a powerful tool for studying the properties of the density matrices.

If ρ is a density matrix having eigenvalues $\lambda_1, \ldots, \lambda_n$, then there is an orthonormal set $\{\boldsymbol{x}_1, \ldots, \boldsymbol{x}_n\}$ of eigenvectors of ρ. Thus, ρ has a spectral representation

$$\rho = \lambda_1 \,|\, \boldsymbol{x}_1\rangle\langle \boldsymbol{x}_1 \,| + \ldots + \lambda_n \,|\, \boldsymbol{x}_n\rangle\langle \boldsymbol{x}_n \,| \,. \tag{8.32}$$

The spectral representation (8.32) in fact tells us that each state (density matrix) is a linear combination of one-dimensional projections, which are alternative notations for quantum states which are defined as unit-length vectors. We will call the states corresponding to one-dimensional projections *vector states* or *pure states.* States that are not pure are referred to as *mixed.*

More information about density matrices can be easily obtained: using the eigenvector basis, it is trivial to see that $\lambda_1 + \ldots + \lambda_n = \mathrm{Tr}(\rho) = 1$. Since ρ is positive, we also see that $\lambda_i = \langle \boldsymbol{x}_i \mid \rho \boldsymbol{x}_i\rangle \geq 0$. Thus, we can say that each mixed state is a *convex combination* of pure states.

Thus, a general density matrix looks very much like a probability distribution of orthogonal pure states. If fact, it is true that each probability distribution $(p_1, \ldots, p_n)$ of mutually orthogonal vector states $\boldsymbol{x}_1, \ldots, \boldsymbol{x}_n$ always defines a density matrix

$$\rho = p_1 \,|\, \boldsymbol{x}_1\rangle\langle \boldsymbol{x}_1 \,| + \ldots + p_n \,|\, \boldsymbol{x}_n\rangle\langle \boldsymbol{x}_n \,|,$$

but the converse is not so clear. For there is one unfortunate feature in the spectral decomposition (8.32). Recall that (8.32) is unique if and only if ρ has distinct eigenvalues. For instance, if $\lambda_1 = \lambda_2 = \lambda$, then it is easy to verify that we can always replace

$$\lambda(|\, \boldsymbol{x}_1\rangle\langle \boldsymbol{x}_1 \,| + |\, \boldsymbol{x}_2\rangle\langle \boldsymbol{x}_2 \,|)$$

with

$$\lambda(|\, \boldsymbol{x}_1'\rangle\langle \boldsymbol{x}_1' \,| + |\, \boldsymbol{x}_2'\rangle\langle \boldsymbol{x}_2' \,|),$$

where

$$\boldsymbol{x}_1' = \boldsymbol{x}_1 \cos\alpha - \boldsymbol{x}_2 \sin\alpha,$$
$$\boldsymbol{x}_2' = \boldsymbol{x}_1 \sin\alpha + \boldsymbol{x}_2 \cos\alpha,$$

and α is a real number (Exercise 5). Therefore, we cannot generally say, without additional information, that a density matrix represents a probability distribution of orthogonal state vectors.

We describe the time evolution of the generalized states briefly at first. Afterwards, in Section 8.4.4, we will study the issue of describing the time evolution in more detail. Here we merely use decomposition (8.32) straightforwardly: for a pure state $|\, \boldsymbol{x}\rangle\langle \boldsymbol{x} \,|$ there is a unitary mapping $U(t)$ such that

$$|\, \boldsymbol{x}(t)\rangle\langle \boldsymbol{x}(t) \,| = |\, U(t)\boldsymbol{x}(0)\rangle\langle U(t)\boldsymbol{x}(0) \,| = U(t) \,|\, \boldsymbol{x}(0)\rangle\langle \boldsymbol{x}(0) \,|\, U(t)^* \tag{8.33}$$

(see Exercise 6). By extending this linearly for mixed states ρ, we have

$$\rho(t) = U(t)\rho(0)U(t)^*,$$

where $U(t)$ is the time-evolution mapping of the system. By using representation $U(t) = \mathrm{e}^{-itH}$, it is quite easy to verify (Exercise 7) that the Schrödinger equation takes the form

$$i\frac{d}{dt}\rho(t) = [H, \rho(t)].$$

8.4.2 Observables and Mixed States

In this section we will demonstrate how to fit the concept of observables into general quantum states. For that purpose, we will first regard an observable as a self-adjoint operator A with spectral representation

$$A = \theta_1 E(\theta_1) + \theta_2 E(\theta_2) + \ldots + \theta_n E(\theta_n),$$

where each $E(\theta_i) = |\boldsymbol{x}_i\rangle\langle\boldsymbol{x}_i|$ is a projection such that the vectors $\boldsymbol{x}_1, \ldots, \boldsymbol{x}_n$ form an orthonormal basis of H_n. If the quantum system is in a state $\boldsymbol{x}$ (pure state $|\boldsymbol{x}\rangle\langle\boldsymbol{x}|$), we can write $\boldsymbol{x} = c_i\boldsymbol{x}_1 + \ldots + c_n\boldsymbol{x}_n$, where $c_i = \langle\boldsymbol{x}_i \mid \boldsymbol{x}\rangle$. Thus, the probability of observing a label θ_i is

$$\begin{aligned}\langle\boldsymbol{x} \mid E(\theta_i)\boldsymbol{x}\rangle &= \Big\langle \sum_{j=1}^{n} \langle\boldsymbol{x}_j \mid \boldsymbol{x}\rangle \boldsymbol{x}_j \mid E(\theta_i)\boldsymbol{x}\Big\rangle \\ &= \sum_{j=1}^{n} \Big\langle \boldsymbol{x}_j \mid E(\theta_i)\langle\boldsymbol{x} \mid \boldsymbol{x}_j\rangle\boldsymbol{x}\Big\rangle \\ &= \sum_{j=1}^{n} \langle\boldsymbol{x}_j \mid E(\theta_i)\,|\boldsymbol{x}\rangle\langle\boldsymbol{x}|\,\boldsymbol{x}_j\rangle \\ &= \mathrm{Tr}\Big(E(\theta_i)\,|\boldsymbol{x}\rangle\langle\boldsymbol{x}|\Big). \end{aligned} \tag{8.34}$$

Equation 8.34 will be the basis for handling the observations on mixed states: Any state T can be expressed as a convex combination

$$T = \lambda_1\,|\boldsymbol{x}_1\rangle\langle\boldsymbol{x}_1| + \ldots + \lambda_n\,|\boldsymbol{x}_n\rangle\langle\boldsymbol{x}_n| \tag{8.35}$$

of pure states $|\boldsymbol{x}_i\rangle\langle\boldsymbol{x}_i|$, and we use the linearity of the trace to join the concept of observables together with the notion of states as self-adjoint mappings. Let T be a state of a quantum system and E an observable which is considered as a projection-valued measure. The probability that a real number (label) in set X is observed is $\mathrm{Tr}(E(X)T) = \mathrm{Tr}(TE(X))$. Notice that, since $\mathrm{Tr}(T) = 1$ and $E(X)$ is a projection, we always have $0 \le \mathrm{Tr}(TE(X)) \le 1$. This can be immediately seen by using the linearity of the trace and the spectral representation (8.35) for T:

$$\mathrm{Tr}(TE(X)) = \sum_{i=1}^{n} \lambda_i \mathrm{Tr}(|\,\boldsymbol{x}_i\rangle\langle \boldsymbol{x}_i\,|\; E(X)) = \sum_{i=1}^{n} \lambda_i \langle \boldsymbol{x}_i \mid E(X)\boldsymbol{x}_i\rangle.$$

Inequalities $0 \le \mathrm{Tr}(TE(X)) \le 1$ now directly follow from the facts that $0 \le \langle \boldsymbol{x}_i \mid E(X)\boldsymbol{x}_i\rangle \le 1$, $\lambda_i \ge 0$, and that $\sum_{i=1}^{n} \lambda_i = 1$.

Example 8.4.1. Let $E = \{E_1, \ldots, E_m\}$ be an observable (a collection of mutually orthogonal subspaces), as in Definition 8.3.1, and P_i the projection onto E_i. If, moreover, $\theta_1, \ldots, \theta_m \in \mathbb{R}$ are the values associated with the subspaces, we can define a projection-valued measure E by $E(\theta_i) = P_i$. Then $A = \theta_1 P_1 + \ldots + \theta_m P_m = \theta_1 E(\theta_1) + \ldots + \theta_m E(\theta_m)$, and the probability that the measurement of observable E will give θ_i as an outcome is given by $\mathrm{Tr}(TP_i) = \mathrm{Tr}(TE(\theta_i))$.

As mentioned above, any state T induces a probability distribution on the set of projections in Hilbert space H_n by $\mu_T(P) = \mathrm{Tr}(TP)$. If $n \ge 3$, the converse also holds. For the statement of the theorem below, recall that $P(H)$ is the set of projections in $L(H)$.

Theorem 8.4.1 (A. Gleason, 1957). *If* $\dim H \ge 3$, *then each probability measure* $\mu : P(H) \to [0,1]$ *satisfying*

1. $\mu(0) = 0$, $\mu(I) = 1$
2. $\mu(\sum_i P_i) = \sum_i \mu(P_i)$

for each sequence of orthogonal projections P_i *is generated via a state* T *(depending only on* μ*) by*

$$\mu(P) = \mathrm{Tr}(TP).$$

The proof of Gleason's theorem can be found in [65], but no simple proof of that theorem is known.

We will demonstrate here how an analogous statement can be reached in a finite-dimensional case, if instead of projections, all self-adjoint mappings are allowed. The proof of the theorem below is from [61].

Theorem 8.4.2. *Let* $L_s(H_n)$ *be the set of self-adjoint operators and* $\mu : L_s(H_n) \to \mathbb{R}$ *a mapping which satisfies*

1. $\mu(I) = 1$
2. $\mu(|\,\boldsymbol{x}\rangle\langle \boldsymbol{x}\,|) \ge 0$ *for each unit-lenght* $\boldsymbol{x} \in H_n$
3. $\mu(\sum_i \alpha_i S_i) = \sum_i \alpha_i \mu(S_i)$ *for each sequence of self-adjoint operators* $S_i \in L_s(H_n)$ *and real numbers* α_i.

Then there is a state T *depending only on* μ *such that*

$$\mu(S) = \mathrm{Tr}(TS)$$

for each self-adjoint operator S.

Before the proof, we will list some simple auxiliary results.

Lemma 8.4.1.

$$\begin{aligned} &c\,|\boldsymbol{x}_r\rangle\langle\boldsymbol{x}_s|+c^*\,|\boldsymbol{x}_r\rangle\langle\boldsymbol{x}_s| \\ &=\mathrm{Re}(c)\big(\,|\boldsymbol{x}_r\rangle\langle\boldsymbol{x}_s|+|\boldsymbol{x}_s\rangle\langle\boldsymbol{x}_r|\,\big)+\mathrm{Im}(c)\big(i\,|\boldsymbol{x}_r\rangle\langle\boldsymbol{x}_s|-i\,|\boldsymbol{x}_s\rangle\langle\boldsymbol{x}_r|\,\big), \end{aligned}$$

where $\mathrm{Re}(c)$ *and* $\mathrm{Im}(c)$ *are the real and imaginary parts of* c*, respectively.*

Proof. Recalling that $\mathrm{Re}(c)=\frac{1}{2}(c+c^*)$ and $\mathrm{Im}(c)=\frac{1}{2i}(c-c^*)$, the statement of the lemma follows by direct calculation. □

Lemma 8.4.2. *Mappings* $B_{rs}=|\boldsymbol{x}_r\rangle\langle\boldsymbol{x}_s|+|\boldsymbol{x}_s\rangle\langle\boldsymbol{x}_r|$ *and* $C_{rs}=i\,|\boldsymbol{x}_r\rangle\langle\boldsymbol{x}_s|$ $-i\,|\boldsymbol{x}_s\rangle\langle\boldsymbol{x}_r|$ *are self-adjoint. Moreover,* $B_{sr}=B_{rs}$ *and* $C_{sr}=-C_{rs}$.

Proof. By direct calculation, recalling Remark 8.2.4. □

Proof (Proof of Theorem 8.4.2). Let $S\in L_s(H_n)$. By fixing an orthonormal basis $\{\boldsymbol{x}_1,\ldots,\boldsymbol{x}_n\}$ of H_n we get a representation (Recall Remark 8.2.5).

$$\begin{aligned} S&=\sum_{r=1}^{n}\sum_{s=1}^{n}\langle\boldsymbol{x}_r\mid S\boldsymbol{x}_s\rangle\,|\boldsymbol{x}_r\rangle\langle\boldsymbol{x}_s| \\ &=\sum_{r=1}^{n}\langle\boldsymbol{x}_r\mid S\boldsymbol{x}_r\rangle\,|\boldsymbol{x}_r\rangle\langle\boldsymbol{x}_r|+\sum_{r\neq s}\langle\boldsymbol{x}_r\mid S\boldsymbol{x}_s\rangle\,|\boldsymbol{x}_r\rangle\langle\boldsymbol{x}_s| \\ &=\sum_{r=1}^{n}\langle\boldsymbol{x}_r\mid S\boldsymbol{x}_r\rangle\,|\boldsymbol{x}_r\rangle\langle\boldsymbol{x}_r| \\ &\quad+\sum_{r<s}\langle\boldsymbol{x}_r\mid S\boldsymbol{x}_s\rangle\,|\boldsymbol{x}_r\rangle\langle\boldsymbol{x}_s|+\sum_{r>s}\langle\boldsymbol{x}_r\mid S\boldsymbol{x}_s\rangle\,|\boldsymbol{x}_r\rangle\langle\boldsymbol{x}_s| \end{aligned}$$

changing the role of r and s in the last term we have that

$$\begin{aligned} S&=\sum_{r=1}^{n}\langle\boldsymbol{x}_r\mid S\boldsymbol{x}_r\rangle\,|\boldsymbol{x}_r\rangle\langle\boldsymbol{x}_r| \\ &\quad+\sum_{r<s}\langle\boldsymbol{x}_r\mid S\boldsymbol{x}_s\rangle\,|\boldsymbol{x}_r\rangle\langle\boldsymbol{x}_s|+\sum_{r<s}\langle\boldsymbol{x}_s\mid S\boldsymbol{x}_r\rangle\,|\boldsymbol{x}_s\rangle\langle\boldsymbol{x}_r| \\ &=\sum_{r=1}^{n}\langle\boldsymbol{x}_r\mid S\boldsymbol{x}_r\rangle\,|\boldsymbol{x}_r\rangle\langle\boldsymbol{x}_r| \\ &\quad+\sum_{r<s}\Big(\langle\boldsymbol{x}_r\mid S\boldsymbol{x}_s\rangle\,|\boldsymbol{x}_r\rangle\langle\boldsymbol{x}_s|+\langle\boldsymbol{x}_r\mid S\boldsymbol{x}_s\rangle^*\,|\boldsymbol{x}_s\rangle\langle\boldsymbol{x}_r|\Big) \end{aligned} \tag{8.36}$$

By applying Lemma 8.4.1 to (8.36), this can be rewritten as

$$\begin{aligned} S = & \sum_{r=1}^{n} \langle \boldsymbol{x}_r \mid S\boldsymbol{x}_r \rangle \, |\boldsymbol{x}_r\rangle\langle \boldsymbol{x}_r | \\ & + \sum_{r<s} \operatorname{Re}\langle \boldsymbol{x}_r \mid S\boldsymbol{x}_s \rangle \big(\, |\boldsymbol{x}_r\rangle\langle \boldsymbol{x}_s | + |\boldsymbol{x}_s\rangle\langle \boldsymbol{x}_r | \, \big) \\ & + \sum_{r<s} \operatorname{Im}\langle \boldsymbol{x}_r \mid S\boldsymbol{x}_s \rangle \big(i \, |\boldsymbol{x}_r\rangle\langle \boldsymbol{x}_s | - i \, |\boldsymbol{x}_s\rangle\langle \boldsymbol{x}_r | \, \big) \end{aligned}$$

By denoting $A_r = |\boldsymbol{x}_r\rangle\langle \boldsymbol{x}_r |$, $B_{rs} = |\boldsymbol{x}_r\rangle\langle \boldsymbol{x}_s | + |\boldsymbol{x}_s\rangle\langle \boldsymbol{x}_r |$, and $C_{rs} = i \, |\boldsymbol{x}_r\rangle\langle \boldsymbol{x}_s | - i \, |\boldsymbol{x}_s\rangle\langle \boldsymbol{x}_r |$ (mappings A_r are clearly self-adjoint and by Lemma 8.4.2 all the mappings B_{rs} and C_{rs} are also self-adjoint), the above equality becomes

$$S = \sum_{r=1}^{n} \langle \boldsymbol{x}_r \mid S\boldsymbol{x}_r \rangle A_r + \sum_{r<s} \operatorname{Re}\langle \boldsymbol{x}_r \mid S\boldsymbol{x}_s \rangle B_{rs} + \sum_{r<s} \operatorname{Im}\langle \boldsymbol{x}_r \mid S\boldsymbol{x}_s \rangle C_{rs}.$$

Because all the coefficients in the above sums are real, and the mappings self-adjoint, the assumption implies that

$$\begin{aligned} \mu(S) = & \sum_{r=1}^{n} \langle \boldsymbol{x}_r \mid S\boldsymbol{x}_r \rangle \mu(A_r) \\ & + \sum_{r<s} \operatorname{Re}\langle \boldsymbol{x}_r \mid S\boldsymbol{x}_s \rangle \mu(B_{rs}) + \sum_{r<s} \operatorname{Im}\langle \boldsymbol{x}_r \mid S\boldsymbol{x}_s \rangle \mu(C_{rs}). \end{aligned}$$

This can be written as

$$\begin{aligned} \mu(S) = & \sum_{r=1}^{n} \langle \boldsymbol{x}_r \mid S\boldsymbol{x}_r \rangle \mu(A_r) \\ & + \sum_{r<s} \operatorname{Re}\langle \boldsymbol{x}_r \mid S\boldsymbol{x}_s \rangle \mu(B_{rs}) + \sum_{r<s} \operatorname{Im}\langle \boldsymbol{x}_r \mid S\boldsymbol{x}_s \rangle \mu(C_{rs}) \\ = & \sum_{r=1}^{n} \langle \boldsymbol{x}_r \mid S\boldsymbol{x}_r \rangle \mu(A_r) \\ & + \sum_{r<s} \frac{1}{2} \big(\langle \boldsymbol{x}_r \mid S\boldsymbol{x}_s \rangle + \langle S\boldsymbol{x}_s \mid \boldsymbol{x}_r \rangle \big) \mu(B_{rs}) \\ & + \sum_{r<s} \frac{1}{2i} \big(\langle \boldsymbol{x}_r \mid S\boldsymbol{x}_s \rangle - \langle S\boldsymbol{x}_r \mid \boldsymbol{x}_r \rangle \big) \mu(C_{rs}). \end{aligned}$$

Equations $B_{rs} = B_{sr}$ and $C_{rs} = -C_{sr}$ imply that

$$\mu(S) = \sum_{r=1}^{n} \langle \boldsymbol{x}_r \mid S\boldsymbol{x}_r \rangle \mu(A_r) + \sum_{r \neq s} \langle \boldsymbol{x}_r \mid S\boldsymbol{x}_s \rangle \Big(\frac{1}{2} \mu(B_{rs}) - \frac{i}{2} \mu(C_{rs}) \Big).$$

Denoting $T_{rr} = \mu(A_r)$, $T_{sr} = \frac{1}{2}\mu(B_{rs}) - \frac{i}{2}\mu(C_{rs})$ we have obviously that $T_{rs}^* = T_{sr}$, so we can define a self-adjoint mapping $T \in L(H_n)$ by

$$T = \sum_{r=1}^{n} \sum_{s=1}^{n} T_{rs} \, |\boldsymbol{x}_r\rangle\langle\boldsymbol{x}_s| \, . \tag{8.37}$$

Equation 8.37 implies directly that $T_{rs} = \langle \boldsymbol{x}_r \mid T\boldsymbol{x}_s \rangle$ and that

$$\begin{aligned}
\mu(S) &= \sum_{r=1}^{n} \sum_{s=1}^{n} T_{sr} \langle \boldsymbol{x}_r \mid S\boldsymbol{x}_s \rangle \\
&= \sum_{r=1}^{n} \sum_{s=1}^{n} \langle \boldsymbol{x}_s \mid T\boldsymbol{x}_r \rangle \langle \boldsymbol{x}_r \mid S\boldsymbol{x}_s \rangle \\
&= \sum_{r=1}^{n} \sum_{s=1}^{n} \langle \boldsymbol{x}_s \, || T\boldsymbol{x}_r \rangle \langle \boldsymbol{x}_r | \, S\boldsymbol{x}_s \rangle \\
&= \sum_{s=1}^{n} \langle \boldsymbol{x}_s \mid T \sum_{r=1}^{n} |\boldsymbol{x}_r\rangle\langle\boldsymbol{x}_r | \, S\boldsymbol{x}_s \rangle \\
&= \sum_{s=1}^{n} \langle \boldsymbol{x}_s \mid TS\boldsymbol{x}_s \rangle = \mathrm{Tr}(TS).
\end{aligned}$$

To finish the proof, we have to show that T is a state. That is, a self-adjoint, unit-trace positive mapping.

We have already noticed that T is self-adjoint. The trace of T can be easily found by summing over the diagonal elements:

$$\mathrm{Tr}(T) = \sum_{r=1}^{n} T_{rr} = \sum_{r=1}^{n} \mu(A_r) = \mu(\sum_{r=1}^{n} A_r) = \mu(I) = 1.$$

For the positivity of T, we notice that if $\boldsymbol{x} \in H_n$ has unit length, then

$$\mu(|\boldsymbol{x}\rangle\langle\boldsymbol{x}|) = \mathrm{Tr}(T \, |\boldsymbol{x}\rangle\langle\boldsymbol{x}|) = \langle \boldsymbol{x} \mid T \, |\boldsymbol{x}\rangle\langle\boldsymbol{x} | \, \boldsymbol{x}\rangle = \langle \boldsymbol{x} \mid T\boldsymbol{x} \rangle.$$

The second-last equality follows from the fact that it is always possible to choose an orthonormal basis containing $\boldsymbol{x}$. Therefore,

$$\langle \boldsymbol{x} \mid T\boldsymbol{x} \rangle = \mu(|\boldsymbol{x}\rangle\langle\boldsymbol{x}|) \geq 0 \tag{8.38}$$

for each unit-length vector $\boldsymbol{x} \in H_n$. If $\boldsymbol{y} \neq \boldsymbol{0}$ is not of unit length, then $\boldsymbol{x} = \frac{1}{||\boldsymbol{y}||}\boldsymbol{y}$ is, and substituting this in (8.38) yields the desired result. □

Remark 8.4.3. Paul Busch introduced a result [22] which states that probability measures defined on *effects*[16] can be expressed analogously to Gleason's theorem. The result of Busch is based on extending the measure μ orginally defined on effects only into a linear functional defined on all self-adjoint operators. The previous theorem can then be applied.

[16] Effect E is an operator in $L_s(H_n)$ satisfying $0 \leq E \leq I$.

8.4.3 Subsystem States

Now we will study the description of compound systems in more detail. Let H_n and H_m be the state spaces of two distinguishable quantum systems. As discussed before, the state space of a compound system is the tensor product $H_n \otimes H_m$. A state of the compound system is a self-adjoint mapping on $H_n \otimes H_m$. Such a state can be comprised of states of subsystems in the following way: if T_1 and T_2 are states (also viewed as self-adjoint mappings) of H_n and H_m, then the tensor product $T_1 \otimes T_2$ defines a self-adjoint mapping on $H_n \otimes H_m$, as is easily seen ($T_1 \otimes T_2$ is defined by

$$(T_1 \otimes T_2)(\boldsymbol{x} \otimes \boldsymbol{y}) = T_1 \boldsymbol{x} \otimes T_2 \boldsymbol{y}$$

for the basis states and extended to $H_n \otimes H_m$ by using linearity requirements). One can also easily verify that the matrix of $T_1 \otimes T_2$ is the tensor product of the matrices of T_1 and T_2. As well, observables of subsystems make up an observable of a compound system: if A_1 and A_2 are observables of subsystems (now seen as unit-trace positive self-adjoint mappings), then $A_1 \otimes A_2$ is a positive, unit-trace self-adjoint mapping in $H_n \otimes H_m$.

The identity mapping $I : H \to H$ is also a positive, unit-trace self-adjoint mapping and, therefore, defines an observable. However, this observable is most uninformative in the following sense: the corresponding projection-valued measure is defined by

$$E_I(X) = \begin{cases} I, \text{ if } 1 \in X, \\ 0, \text{ otherwise,} \end{cases}$$

so the probability distribution associated with E_I is trivial: $\text{Tr}(TE_i(X))$ is 1 if $1 \in X$, 0 otherwise. Notice also that I has only one eigenvalue 1, but all the vectors of H are eigenvectors of I. This strongly reflects on the corresponding decomposition of H into orthogonal subspaces: the decomposition has only one component, H itself. This corresponds to the idea that, by measuring observable I, we are not observing any nontrivial property. In fact, if T_1 and T_2 are the states of H_n and H_m and A_1 and A_2 are some observables, then it is easy to see that

$$\text{Tr}\big((T_1 \otimes T_2)(A_1 \otimes I)\big) = \text{Tr}(T_1 A_1),$$
$$\text{Tr}\big((T_1 \otimes T_2)(I \otimes A_2)\big) = \text{Tr}(T_2 A_2).$$

That is, observing the compound observable $A_1 \otimes I$ (resp., $I \otimes A_2$) corresponds to observing only the first (resp., second) system. Based on this idea, we will define the substates of a compound system.

Definition 8.4.2. *Let T be a state of a compound system with state space $H_n \otimes H_m$. The states of the subsystems are unit-trace positive self-adjoint mappings $T_1 : H_n \to H_n$ and $T_2 : H_m \to H_m$ such that for any observables A_1 and A_2,*

$$\mathrm{Tr}(T_1 A_1) = \mathrm{Tr}(T(A_1 \otimes I)), \tag{8.39}$$

$$\mathrm{Tr}(T_2 A_2) = \mathrm{Tr}(T(I \otimes A_2)). \tag{8.40}$$

We say that T_1 is obtained by tracing over H_m, and T_2 by tracing over H_n.

We can, of course, be skeptical about the existence and uniqueness of T_1 and T_2, but, fortunately, they are always uniquely determined.

Theorem 8.4.3. *For each state T there are unique states T_1 and T_2 satisfying (8.39) and (8.40) for all observables A_1 and A_2.*

Proof. We will show only that T_1 exists and is unique; the proof for T_2 is symmetrical. We will first search for a mapping T_1 that will satisfy

$$\mathrm{Tr}(T_1 A) = \mathrm{Tr}(T(A \otimes I)) \tag{8.41}$$

for *any* linear mapping $A : H_n \to H_n$. For that purpose, we fix orthonormal bases $\{\boldsymbol{x}_1, \ldots, \boldsymbol{x}_n\}$ and $\{\boldsymbol{y}_1, \ldots, \boldsymbol{y}_m\}$ of H_n and H_m, respectively. Using these bases, (8.41) can be rewritten as

$$\begin{aligned}\sum_{i=1}^{n} \langle \boldsymbol{x}_i \mid T_1 A \boldsymbol{x}_i \rangle &= \sum_{i=1}^{n} \sum_{j=1}^{m} \langle \boldsymbol{x}_i \otimes \boldsymbol{y}_j \mid T(A \otimes I)(\boldsymbol{x}_i \otimes \boldsymbol{y}_j) \rangle \\ &= \sum_{i=1}^{n} \sum_{j=1}^{m} \langle \boldsymbol{x}_i \otimes \boldsymbol{y}_j \mid T(A\boldsymbol{x}_i \otimes \boldsymbol{y}_j) \rangle. \end{aligned} \tag{8.42}$$

Recall that T_1 can be represented as

$$T_1 = \sum_{i=1}^{n} \sum_{j=1}^{n} c_{ij} \, |\boldsymbol{x}_i\rangle\langle \boldsymbol{x}_j|, \tag{8.43}$$

where c_{ij} are complex numbers (Remark 8.2.5).

By choosing $A = |\boldsymbol{x}_k\rangle\langle \boldsymbol{x}_l|$ and substituting (8.43) in (8.42), we obtain

$$c_{lk} = \sum_{j=1}^{m} \langle \boldsymbol{x}_l \otimes \boldsymbol{y}_j \mid T(\boldsymbol{x}_k \otimes \boldsymbol{y}_j) \rangle,$$

which gives

$$T_1 = \sum_{i=1}^{n} \sum_{j=1}^{n} \sum_{k=1}^{m} \langle \boldsymbol{x}_i \otimes \boldsymbol{y}_k \mid T(\boldsymbol{x}_j \otimes \boldsymbol{y}_k) \rangle \, |\boldsymbol{x}_i\rangle\langle \boldsymbol{x}_j| \tag{8.44}$$

as a candidate for T_1. In fact, mapping (8.44) is self-adjoint:

$$\begin{aligned} T_1^* &= \sum_{i=1}^{n} \sum_{j=1}^{n} \sum_{k=1}^{m} \langle T(\boldsymbol{x}_j \otimes \boldsymbol{y}_k) \mid \boldsymbol{x}_i \otimes \boldsymbol{y}_k \rangle \, |\boldsymbol{x}_j\rangle\langle \boldsymbol{x}_i| \\ &= \sum_{i=1}^{n} \sum_{j=1}^{n} \sum_{k=1}^{m} \langle \boldsymbol{x}_j \otimes \boldsymbol{y}_k \mid T(\boldsymbol{x}_i \otimes \boldsymbol{y}_k) \rangle \, |\boldsymbol{x}_j\rangle\langle \boldsymbol{x}_i| \\ &= T_1. \end{aligned}$$

By choosing $A = I$, we also see that $\mathrm{Tr}(T_1) = 1$. What about the positivity of T_1? Analogously to (8.34),

$$\langle \boldsymbol{x} \mid T_1 \boldsymbol{x} \rangle = \mathrm{Tr}(T_1 \mid \boldsymbol{x} \rangle \langle \boldsymbol{x} \mid) = \mathrm{Tr}(T(\mid \boldsymbol{x} \rangle \langle \boldsymbol{x} \mid \otimes I)) \geq 0,$$

so T_1 is also positive. This proves the existence of the required state T_1.

Assume then, on the contrary, that there is another state T_1' such that $\mathrm{Tr}(T_1 A) = \mathrm{Tr}(T_1' A)$ for each self-adjoint A. By choosing A as a projection $\mid \boldsymbol{x} \rangle \langle \boldsymbol{x} \mid$, we have that $\langle \boldsymbol{x} \mid (T_1 - T_1')\boldsymbol{x} \rangle = 0$ for any unit-length $\boldsymbol{x}$. Mapping $T_1 - T_1'$ is also self-adjoint, so there is an orthonormal basis of H_n consisting of eigenvectors of $T_1 - T_1'$. If all the eigenvalues of $T_1 - T_1'$ are 0, then $T_1 - T_1' = 0$, and the proof is complete. In the opposite case, there is a nonzero eigenvalue λ_1 of $T_1 - T_1'$ and a unit-length eigenvector $\boldsymbol{x}_1$ belonging to λ_1. Then,

$$\langle \boldsymbol{x}_1 \mid (T_1 - T_1')\boldsymbol{x}_1 \rangle = \lambda_1 \langle \boldsymbol{x}_1 \mid \boldsymbol{x}_1 \rangle \neq 0,$$

a contradiction. □

Let us write $T_1 = \mathrm{Tr}_{H_m}(T)$ for the state obtained by tracing over H_m and, similarly, $T_2 = \mathrm{Tr}_{H_n}(T)$ for the state that we get by tracing over H_n. By collecting all facts in the previous proof, we see that

$$T_1 = \sum_{i=1}^{n} \sum_{j=1}^{n} \sum_{k=1}^{m} \langle \boldsymbol{x}_i \otimes \boldsymbol{y}_k \mid T(\boldsymbol{x}_j \otimes \boldsymbol{y}_k) \rangle \mid \boldsymbol{x}_i \rangle \langle \boldsymbol{x}_j \mid . \tag{8.45}$$

Similarly,

$$T_2 = \sum_{i=1}^{m} \sum_{j=1}^{m} \sum_{k=1}^{n} \langle \boldsymbol{x}_k \otimes \boldsymbol{y}_i \mid T(\boldsymbol{x}_k \otimes \boldsymbol{y}_j) \rangle \mid \boldsymbol{y}_i \rangle \langle \boldsymbol{y}_j \mid . \tag{8.46}$$

It is easy to see that the tracing-over operation is linear: $\mathrm{Tr}_{H_n}(\alpha S + \beta T) = \alpha \mathrm{Tr}_{H_n}(S) + \beta \mathrm{Tr}_{H_n}(T)$.

Example 8.4.2. Let the notation be as in the previous lemma. If S is a state of the compound system $H_n \otimes H_m$, then S is, in particular, a linear mapping $H_n \otimes H_m \to H_n \otimes H_m$. Therefore, S can be uniquely represented as

$$S = \sum_{r=1}^{n} \sum_{t=1}^{m} \sum_{s=1}^{n} \sum_{u=1}^{m} s_{rstu} \mid \boldsymbol{x}_r \otimes \boldsymbol{y}_t \rangle \langle \boldsymbol{x}_s \otimes \boldsymbol{y}_u \mid . \tag{8.47}$$

It is easy to verify that $\mid \boldsymbol{x}_r \otimes \boldsymbol{y}_t \rangle \langle \boldsymbol{x}_s \otimes \boldsymbol{y}_u \mid = \mid \boldsymbol{x}_r \rangle \langle \boldsymbol{x}_s \mid \otimes \mid \boldsymbol{y}_t \rangle \langle \boldsymbol{y}_u \mid$, so we can write

$$\begin{aligned} S &= \sum_{r=1}^{n} \sum_{s=1}^{n} \mid \boldsymbol{x}_r \rangle \langle \boldsymbol{x}_s \mid \otimes \sum_{t=1}^{m} \sum_{u=1}^{m} s_{rstu} \mid \boldsymbol{y}_t \rangle \langle \boldsymbol{y}_u \mid \\ &= \sum_{r=1}^{n} \sum_{s=1}^{n} \mid \boldsymbol{x}_r \rangle \langle \boldsymbol{x}_s \mid \otimes S_{rs}, \end{aligned} \tag{8.48}$$

where each $S_{rs} \in L(H_m)$ (recall that the notation $L(H_m)$ stands for the linear mappings $H_m \to H_m$). It is worth noticing that the latest expression for S is actually a decomposition of an $nm \times nm$ matrix (8.47) into an $n \times n$ block matrix having $m \times m$ matrices S_{rs} as entries.

Substituting (8.48) in (8.45), we get, after some calculation, that

$$S_2 = \sum_{r=1}^{n}\sum_{i=1}^{m}\sum_{j=1}^{m} \langle \boldsymbol{y}_i \mid S_{rr}\boldsymbol{y}_j\rangle \mid \boldsymbol{y}_i\rangle\langle \boldsymbol{y}_j \mid = \sum_{r=1}^{n} S_{rr}.$$

The above expression can be interpreted as follows: if the density matrix corresponding to the state S is viewed as an $n \times n$ block matrix with $m \times m$ blocks S_{rs}, then the density matrix corresponding to the state of system S_2 can be obtained by summing together all the diagonal blocks S_{rr}.

Example 8.4.3. Let us investigate the compound system $H_2 \otimes H_2$ of two qubits. We choose $\{|00\rangle, |01\rangle, |10\rangle, |11\rangle\}$ as the orthonormal basis of $H_2 \otimes H_2$. Consider the (pure) state $\boldsymbol{y} = \frac{1}{\sqrt{2}}(|00\rangle + |11\rangle)$ of the compound system.

By using (8.45), we see that the states of the subsystems are

$$S' = \frac{1}{2} \mid 0\rangle\langle 0 \mid + \frac{1}{2} \mid 1\rangle\langle 1 \mid .$$

By using the familiar coordinate representation $|00\rangle = (1,0,0,0)^T$, $|01\rangle = (0,1,0,0)^T$, $|10\rangle = (0,0,1,0)^T$, and $|11\rangle = (0,0,0,1)^T$, the coordinate representation of $\boldsymbol{y}$ is $(\frac{1}{\sqrt{2}}, 0, 0, \frac{1}{\sqrt{2}})^T$ and it is easy to see that the density matrix corresponding to the pure state $|\boldsymbol{y}\rangle\langle\boldsymbol{y}|$ is

$$M_{|\boldsymbol{y}\rangle\langle\boldsymbol{y}|} = \frac{1}{2}\begin{pmatrix} 1&0&0&1\\ 0&0&0&0\\ 0&0&0&0\\ 1&0&0&1 \end{pmatrix}$$

and the density matrices correspoding to states S' are

$$M_{S'} = \frac{1}{2}\begin{pmatrix} 1&0\\ 0&1 \end{pmatrix},$$

which agrees with the previous example. On the other hand, if we try to reconstruct the state of $H_2 \otimes H_2$ from states S', we find out that

$$M_{S'} \otimes M_{S'} = \frac{1}{4}\begin{pmatrix} 1&0&0&0\\ 0&1&0&0\\ 0&0&1&0\\ 0&0&0&1 \end{pmatrix},$$

which is different from $M_{|\boldsymbol{y}\rangle\langle\boldsymbol{y}|}$. This demonstrates that *subsystem states S' obtained by tracing over are not enough to determine the state of the compound system.*

Even though the state of the compound system perfectly determines the subsystem states, there is no way to obtain the whole system state from the partial systems without additional information.

Example 8.4.4. Let the notation be as before. If $\boldsymbol{z} \in H_n \otimes H_m$ is a unit-length vector, then $|\,\boldsymbol{z}\rangle\langle\boldsymbol{z}\,|$ corresponds to a pure state of system $H_n \otimes H_m$. For each pair $(\boldsymbol{x}, \boldsymbol{z}) \in H_n \times H_n \otimes H_m$, there exists a unique vector $\boldsymbol{y} \in H_m$ such that

$$\langle \boldsymbol{y}' \mid \boldsymbol{y}\rangle = \langle \boldsymbol{x} \otimes \boldsymbol{y}' \mid \boldsymbol{z}\rangle$$

for each $\boldsymbol{y}' \in H_m$. In fact, if $\boldsymbol{y}_1$ and $\boldsymbol{y}_2$ were two such vectors, then $\langle \boldsymbol{y}' \mid \boldsymbol{y}_1 - \boldsymbol{y}_2\rangle = 0$ for each $\boldsymbol{y}' \in H_m$, and $\boldsymbol{y}_1 = \boldsymbol{y}_2$ follows immediately. On the other hand,

$$\boldsymbol{y} = \sum_{k=1}^{m} \langle \boldsymbol{x} \otimes \boldsymbol{y}_k \mid \boldsymbol{z}\rangle \boldsymbol{y}_k$$

clearly satisfies the condition required. We define a mapping $(\cdot,\cdot) : H_n \times H_n \otimes H_m \to H_m$ by

$$(\boldsymbol{x}, \boldsymbol{z}) = \sum_{k=1}^{m} \langle \boldsymbol{x} \otimes \boldsymbol{y}_k \mid \boldsymbol{z}\rangle \boldsymbol{y}_k.$$

Clearly $(\cdot,\cdot)$ has properties that resemble the inner product, such as linearity with respect to the second component and antilinearity[17] with respect to the first component.

Now, substituting $S = |\,\boldsymbol{z}\rangle\langle\boldsymbol{z}\,|$ in (8.45), we see that

$$S_2 = \sum_{k=1}^{n} |\,(\boldsymbol{x}_k, \boldsymbol{z})\rangle\langle(\boldsymbol{x}_k, \boldsymbol{z})\,|\,. \tag{8.49}$$

Comparing (8.49) with the equation

$$\mathrm{Tr}(|\,\boldsymbol{z}\rangle\langle\boldsymbol{z}\,|) = \sum_{k=1}^{n} \langle \boldsymbol{x}_k \,||\,\boldsymbol{z}\rangle\langle\boldsymbol{z}\,|\,\boldsymbol{x}_k\rangle = \sum_{k=1}^{n} \langle \boldsymbol{x}_k \mid \boldsymbol{z}\rangle\langle\boldsymbol{z} \mid \boldsymbol{x}_k\rangle$$

somehow explains the name "tracing over". Notice also that if

$$\boldsymbol{z} = \sum_{i=1}^{n}\sum_{j=1}^{m} c_{ij}\boldsymbol{x}_i \otimes \boldsymbol{y}_j = \sum_{i=1}^{n} \boldsymbol{x}_i \otimes \sum_{j=1}^{m} c_{ij}\boldsymbol{y}_j,$$

then clearly $(\boldsymbol{x}_k, \boldsymbol{z}) = \sum_{j=1}^{m} c_{kj}\boldsymbol{y}_j$, and S_2 becomes

$$S_2 = \sum_{k=1}^{n} |\sum_{j=1}^{m} c_{kj}\boldsymbol{y}_j\rangle\langle\sum_{j=1}^{m} c_{kj}\boldsymbol{y}_j\,|\,.$$

[17] Here antilinearity, as usual, means that $(\alpha_1\boldsymbol{x}_1 + \alpha_2\boldsymbol{x}_2, \boldsymbol{z}) = \alpha_1^*(\boldsymbol{x}_1, \boldsymbol{z}) + \alpha_2^*(\boldsymbol{x}_2, \boldsymbol{z})$.

Theorem 8.4.4 (Schmidt decomposition). *Assume that* $n \le m$. *Let* $\boldsymbol{z} \in H_n \otimes H_m$ *be a unit-length vector and* $T_1 = \mathrm{Tr}_{H_m}(|\,\boldsymbol{z}\rangle\langle\boldsymbol{z}\,|)$ *and* $T_2 = \mathrm{Tr}_{H_n}(|\,\boldsymbol{z}\rangle\langle\boldsymbol{z}\,|)$ *the subsystem states in* $L(H_n)$ *and* $L(H_m)$, *respectively. Let also* $\{\boldsymbol{x}_1, \ldots, \boldsymbol{x}_n\}$ *be an orthonormal basis of* H_n *consisting of eigenvectors of* T_1 *and*

$$T_1 = \lambda_1\,|\,\boldsymbol{x}_1\rangle\langle\boldsymbol{x}_1\,| + \ldots + \lambda_n\,|\,\boldsymbol{x}_n\rangle\langle\boldsymbol{x}_n\,|$$

a spectral representation of T_1. *Then,*

$$\boldsymbol{z} = \sqrt{\lambda_1}\boldsymbol{x}_1 \otimes \boldsymbol{y}_1 + \ldots + \sqrt{\lambda_n}\boldsymbol{x}_n \otimes \boldsymbol{y}_n,$$

where $\{\boldsymbol{y}_i \mid \lambda_i \neq 0\}$ *is a set of orthonormal eigenvectors of* T_2 *(not necessarily a basis of* H_m*).*

Proof. Let $\{\boldsymbol{b}_1, \ldots, \boldsymbol{b}_m\}$ be an orthonormal basis of H_m. Then, vectors $\boldsymbol{x}_i \otimes \boldsymbol{b}_j$ form an orthonormal basis of $H_n \otimes H_m$, and $\boldsymbol{z}$ can be represented as

$$\boldsymbol{z} = \sum_{i=1}^{n}\sum_{j=1}^{m} c_{ij}\boldsymbol{x}_i \otimes \boldsymbol{b}_j = \sum_{i=1}^{n} \boldsymbol{x}_i \otimes \boldsymbol{y}_i', \tag{8.50}$$

where

$$\boldsymbol{y}_i' = \sum_{j=1}^{m} c_{ij}\boldsymbol{b}_j.$$

On the other hand, we can write

$$\boldsymbol{z} = \sum_{j=1}^{m} \boldsymbol{x}_j' \otimes \boldsymbol{b}_j, \tag{8.51}$$

where

$$\boldsymbol{x}_j' = \sum_{i=1}^{n} c_{ij}\boldsymbol{x}_i.$$

By applying the previous example to (8.51), we get

$$\begin{aligned} T_1 = \mathrm{Tr}_{H_m}(|\,\boldsymbol{z}\rangle\langle\boldsymbol{z}\,|) &= \sum_{k=1}^{m} |\,\boldsymbol{x}_k'\rangle\langle\boldsymbol{x}_k'\,| = \sum_{k=1}^{m} |\sum_{i=1}^{n} c_{ik}\boldsymbol{x}_i\rangle\langle\sum_{j=1}^{n} c_{jk}\boldsymbol{x}_j\,| \\ &= \sum_{k=1}^{m}\sum_{i=1}^{n}\sum_{j=1}^{n} c_{ik}c_{jk}^*\,|\,\boldsymbol{x}_i\rangle\langle\boldsymbol{x}_j\,|\,. \end{aligned} \tag{8.52}$$

On the other hand,

$$\begin{aligned} \langle\boldsymbol{y}_j' \mid \boldsymbol{y}_i'\rangle &= \langle\sum_{k=1}^{m} c_{jk}\boldsymbol{b}_k \mid \sum_{l=1}^{m} c_{il}\boldsymbol{b}_l\rangle \\ &= \sum_{k=1}^{m}\sum_{l=1}^{m} c_{jk}^*c_{il}\langle\boldsymbol{b}_k \mid \boldsymbol{b}_l\rangle = \sum_{k=1}^{m} c_{jk}^*c_{ik}, \end{aligned}$$

so (8.52) can be rewritten as

$$T_1 = \sum_{i=1}^{n}\sum_{j=1}^{n} \langle \boldsymbol{y}'_j \mid \boldsymbol{y}'_i\rangle \, |\boldsymbol{x}_i\rangle\langle \boldsymbol{x}_j| \, . \tag{8.53}$$

Now that T_1 also possesses the spectral representation

$$T_1 = \sum_{i=1}^{n} \lambda_i \, |\boldsymbol{x}_i\rangle\langle \boldsymbol{x}_i|$$

and representation (8.53) is unique, we must conclude that

$$\langle \boldsymbol{y}'_j \mid \boldsymbol{y}'_i\rangle = \begin{cases} \lambda_i, \text{ if } j = i \\ 0 \quad \text{otherwise.} \end{cases}$$

Substituting $\boldsymbol{y}'_i = \sqrt{\lambda_i}\boldsymbol{y}_i$ to (8.50) gives the desired decomposition, and clearly $\{\boldsymbol{y}_i \mid \lambda_i \neq 0\}$ is an orthonormal set. Applying Example 8.4.3 to decomposition

$$\boldsymbol{z} = \sum_{i=1}^{n} \sqrt{\lambda_i}\boldsymbol{x} \otimes \boldsymbol{y}_i$$

now directly yields

$$T_2 = \sum_{i=1}^{n} \lambda_i \, |\boldsymbol{y}_i\rangle\langle \boldsymbol{y}_i|,$$

which means that $\boldsymbol{y}_1, \ldots, \boldsymbol{y}_n$ are eigenvectors of T_2 belonging to eigenvalues $\lambda_1, \ldots, \lambda_n$. □

Lemma 8.4.3 (Purification). *Let $T \in L(H_n)$ be an arbitrary state of a quantum system. There exists an integer m and a unit-length vector $\boldsymbol{z} \in H_n \otimes H_m$ such that $T = \mathrm{Tr}_{H_m}(|\boldsymbol{z}\rangle\langle \boldsymbol{z}|)$.*

Proof. Let

$$T = \lambda_1 \, |\boldsymbol{x}_1\rangle\langle \boldsymbol{x}_1| + \ldots + \lambda_n \, |\boldsymbol{x}_n\rangle\langle \boldsymbol{x}_n|$$

be a spectral representation of T. The previous theorem and Example 8.4.3 give us a clue how to construct $\boldsymbol{z}$: we choose $n = m$ and let $\boldsymbol{y}_1, \ldots, \boldsymbol{y}_n$ be an orthonormal basis of another copy of H_n. We define

$$\boldsymbol{z} = \sqrt{\lambda_1}\boldsymbol{x}_1 \otimes \boldsymbol{y}_1 + \ldots + \sqrt{\lambda_n}\boldsymbol{x}_n \otimes \boldsymbol{y}_n.$$

Notice that since T is a positive mapping, all eigenvalues λ_i are nonnegative, and the square roots can be therefore defined. The length of $\boldsymbol{z}$ can also be easily computed:

$$||\boldsymbol{z}|| = (\sqrt{\lambda_1})^2 + \ldots + (\sqrt{\lambda_n})^2 = \lambda_1 + \ldots + \lambda_n = 1,$$

because T has unit trace.

Now $\boldsymbol{z}$ fits in the previous example by setting

$$c_{ij} = \begin{cases} \sqrt{\lambda_i} & \text{if } i = j \\ 0 & \text{otherwise.} \end{cases}$$

By applying the result of the previous example, we see directly that

$$\mathrm{Tr}_{H_m}(|\,\boldsymbol{z}\rangle\langle\boldsymbol{z}\,|) = \sum_{k=1}^{n} |\,\sqrt{\lambda_k}\boldsymbol{x}_k\rangle\langle\sqrt{\lambda_k}\boldsymbol{x}_k\,| = \sum_{k=1}^{n} \lambda_k \,|\,\boldsymbol{x}_k\rangle\langle\boldsymbol{x}_k\,| = T.$$

□

Remark 8.4.4. The above lemma states that any state of a quantum system can be interpreted as a result of tracing over a pure state of a larger system. If T is a mixed state, a pure state $|\,\boldsymbol{z}\rangle\langle\boldsymbol{z}\,|$ such that $T = \mathrm{Tr}_{H_m}(|\,\boldsymbol{z}\rangle\langle\boldsymbol{z}\,|)$ is called a *purification* of T.

8.4.4 More on Time Evolution

We already know how to interprete abstract mathematical concepts such as state (positive, unit-trace self-adjoint operator) and observable (can be seen as a self-adjoint operator). Since these concepts refer to physical objects, it is equally important to know how the description of a system changes in time. In other words, we should know how to describe the *dynamic evolution* of a system. We have already outlined in Section 8.3.1 how time evolution can be described by the state vector formalism. Here, we will discuss the time evolution of the general formalism in more detail.

In the so-called *Schrödinger picture*, the state of a system may change in time but the observables remain, whereas in the *Heisenberg picture* the observables change. It turns out that both pictures are mathematically equivalent, and here we will adopt the Schrödinger picture.

We will also adopt, as in Section 8.3.1, the *causality principle*, which tells us that the state of a system at some time t can be recovered from the state at a given time $t = 0$. Mathematically speaking, this means that if the state of the system H_n at $t = 0$ is S_0, there exists a function $V_t : L(H_n) \to L(H_n)$ such that the state at a given time t is $S_t = V_t(S_0)$.

It follows that each V_t should preserve self-adjointness, trace, and positivity. A requirement potentially subject to more criticism is that each V_t must be *linear*.

It is also quite common to require that each V_t is *completely positive*. To define the concept of a completely positive linear mapping $L(H_n) \to L(H_n)$, we consider first some basic properties of tensor products.

For some natural number $m \geq 1$, let H_m be a Hilbert space with orthonormal basis $\{\boldsymbol{y}_1, \ldots, \boldsymbol{y}_m\}$. If $\{\boldsymbol{x}_1, \ldots, \boldsymbol{x}_n\}$ is also an orthonormal basis of H_n, then

$$\{\boldsymbol{y}_i \otimes \boldsymbol{x}_j \mid i \in \{1, \ldots, m\}, j \in \{1, \ldots, n\}\}$$

is a natural choice for an orthonormal basis of $H_m \otimes H_n$. We will keep this notation throughout this section.

Each linear mapping $W \in L(H_m \otimes H_n)$ can be uniquely represented as

$$W = \sum_{r=1}^{m} \sum_{s=1}^{m} |\boldsymbol{y}_r\rangle\langle \boldsymbol{y}_s| \otimes A_{rs}, \tag{8.54}$$

where each $A_{rs} \in L(H_n)$ (see Example 8.4.2). Whenever W is represented as in (8.54), then knowing each A_{rs} uniquely determines W and vice versa, since

$$\langle \boldsymbol{y}_k \otimes \boldsymbol{y}_i \mid W(\boldsymbol{y}_l \otimes \boldsymbol{x}_j)\rangle = \langle \boldsymbol{x}_i \mid A_{kl}\boldsymbol{x}_j\rangle,$$

and any linear mapping in $A_{kl} \in L(H_n)$ is completely determined by the values $\langle \boldsymbol{x}_i \mid A_{kl}\boldsymbol{x}_j\rangle$. In fact, this is already quite clear from the fact that (8.54) actually represents an $mn \times mn$ matrix W as an $m \times m$ matrix with $n \times n$ matrices as entries.

Now, if $V : L(H_n) \to L(H_n)$ is a linear mapping and W is as in (8.54), we define another mapping $I_m \otimes W : L(H_m \otimes H_n) \to L(H_m \otimes H_n)$ by

$$(I_m \otimes V)(W) = \sum_{r=1}^{m} \sum_{s=1}^{m} |\boldsymbol{y}_r\rangle\langle \boldsymbol{y}_s| \otimes V(A_{rs}).$$

Definition 8.4.3. *Mapping $V : L(H_n) \to L(H_n)$ is completely positive if for any $m \geq 1$, the extended mapping $I_m \otimes V$ preserves positivity.*

Remark 8.4.5. The intuitive meaning of the notion "completely positive" is the following: instead of merely regarding T as a state of the quantum system H_n, we could as well introduce an "environment" H_m and regard $I_m \otimes T$ as the state of the compound system $H_m \otimes H_n$. The requirement that V should be completely positive means that the time evolution on a compound system must preserve positivity as well.

Trace-preserving completely positive mappings, which now will be assumed as general time evolution operations of quantum systems, are also known as *superoperators*. If the time evolution of a quantum system can be described by unitary mappings, we call the system *closed*. Otherwise, a quantum system is said to be *open*.

Because of their conceptual importance, we will devote the next section to the study of completely positive mappings.

8.4.5 Representation Theorems

We will derive two representation theorems of completely positive mappings. For that purpose, we begin with some auxiliary results.

Lemma 8.4.4. *A positive mapping $A \in L(H_n)$ is self-adjoint.*

Proof. The positivity of A means that $\langle \boldsymbol{x} \mid A\boldsymbol{x} \rangle \geq 0$ for each $\boldsymbol{x} \in H_n$. Specifically, $\langle \boldsymbol{x} \mid A\boldsymbol{x} \rangle$ is real, hence $\langle \boldsymbol{x} \mid A\boldsymbol{x} \rangle = \langle A\boldsymbol{x} \mid \boldsymbol{x} \rangle$ for each $\boldsymbol{x} \in H_n$. On the other hand, also $\langle \boldsymbol{x} \mid A\boldsymbol{x} \rangle = \langle A^*\boldsymbol{x} \mid \boldsymbol{x} \rangle$ for each $\boldsymbol{x} \in H_n$. Writing $B = i(A - A^*)$ we have that $\langle B\boldsymbol{x} \mid \boldsymbol{x} \rangle = 0$ for any $\boldsymbol{x} \in H_n$. Since B is self-adjoint, there exists an orthonormal basis $\{\boldsymbol{x}_1, \ldots, \boldsymbol{x}_n\}$ consisting of eigenvectors of B and

$$B = \lambda_1 \left|\boldsymbol{x}_1\right\rangle\left\langle\boldsymbol{x}_1\right| + \ldots + \lambda_n \left|\boldsymbol{x}_n\right\rangle\left\langle\boldsymbol{x}_n\right|;$$

see spectral representation (8.16). But then

$$0 = \langle B\boldsymbol{x}_i \mid \boldsymbol{x}_i \rangle = \lambda_i^* \langle \boldsymbol{x}_i \mid \boldsymbol{x}_i \rangle = \lambda_i^*$$

for each $i \in \{1, \ldots, n\}$, and therefore $B = 0$. Equality $A = A^*$ follows immediately. □

Lemma 8.4.5. *A mapping $A \in L(H_n)$ is positive if and only if there is an orthonormal basis $\{\boldsymbol{x}_1, \ldots, \boldsymbol{x}_n\}$ of H_n and nonnegative numbers $\lambda_1, \ldots, \lambda_n$ such that*

$$A = \lambda_1 \left|\boldsymbol{x}_1\right\rangle\left\langle\boldsymbol{x}_1\right| + \ldots + \lambda_n \left|\boldsymbol{x}_n\right\rangle\left\langle\boldsymbol{x}_n\right|. \tag{8.55}$$

Proof. Assume that A has representation (8.55). Each $\boldsymbol{x} \in H_n$ can be written as

$$\boldsymbol{x} = \alpha_1 \boldsymbol{x}_1 + \ldots + \alpha_n \boldsymbol{x}_n,$$

and then $\langle \boldsymbol{x} \mid A\boldsymbol{x} \rangle = \lambda_1 \left|\alpha_1\right|^2 + \ldots + \lambda_n \left|\alpha_n\right|^2 \geq 0$, hence A is positive.

On the other hand, if A is positive, then A is also self-adjoint by the previous lemma. Using the spectral representation, we see that there is an orthonormal $\{\boldsymbol{x}_1, \ldots, \boldsymbol{x}_n\}$ basis of H_n such that

$$A = \lambda_1 \left|\boldsymbol{x}_1\right\rangle\left\langle\boldsymbol{x}_1\right| + \ldots + \lambda_n \left|\boldsymbol{x}_n\right\rangle\left\langle\boldsymbol{x}_n\right|$$

for some real numbers $\lambda_1, \ldots, \lambda_n$. Since A is positive,

$$0 \leq \langle \boldsymbol{x}_i \mid A\boldsymbol{x}_i \rangle = \langle \boldsymbol{x}_i \mid \lambda_i \boldsymbol{x}_i \rangle = \lambda_i$$

for each i. □

The proof of the following theorem, the first structure theorem, is due to [25].

Theorem 8.4.5. *Linear mapping $V : L(H_n) \to L(H_n)$ is completely positive if and only if there exists n^2 mappings $V_i \in L(H_n)$ such that*

$$V(A) = \sum_{i=1}^{n^2} V_i A V_i^* \tag{8.56}$$

for each $A \in L(H_n)$.

Remark 8.4.6. In the quantum computation community, (8.56) is frequently referred as to the *Kraus representation* of a completely positive mapping.

Proof. If V has representation (8.56), we should show that whenever

$$W = \sum_{r=1}^{m}\sum_{s=1}^{m} |\, \boldsymbol{y}_r\rangle\langle \boldsymbol{y}_s \,| \otimes A_{rs}$$

is positive, then also

$$(I_m \otimes V)(W) = \sum_{r=1}^{m}\sum_{s=1}^{m} |\, \boldsymbol{y}_r\rangle\langle \boldsymbol{y}_s \,| \otimes \sum_{i=1}^{n^2} V_i A_{rs} V_i^*$$

is positive. This is done by straightforward computation: Any $\boldsymbol{z} \in H_m \otimes H_n$ can be represented in the form

$$\boldsymbol{z} = \sum_{\alpha=1}^{N} \boldsymbol{y}_\alpha \otimes \boldsymbol{x}_\alpha,$$

where $N \le n^2$ and each $\boldsymbol{y}_\alpha \in H_m$ and $\boldsymbol{x}_\alpha \in H_n$. Writing

$$\boldsymbol{z}_i = \sum_{\alpha=1}^{N} \boldsymbol{y}_\alpha \otimes V_i^* \boldsymbol{x}_\alpha,$$

it is easy to check that

$$\langle \boldsymbol{z} \mid (I_m \otimes V)(W)\boldsymbol{z}\rangle = \sum_{i=1}^{n^2} \langle \boldsymbol{z}_i \mid W \boldsymbol{z}_i\rangle \ge 0,$$

because W is positive.

For the other direction, let V be a completely positive mapping. Then particularly, $I_n \otimes V$ should be positive. For clarity, we will denote the basis of one copy of H_n by $\{\boldsymbol{y}_1, \ldots, \boldsymbol{y}_n\}$ and that of the other copy by $\{\boldsymbol{x}_1, \ldots, \boldsymbol{x}_n\}$. Mapping $W \in L(H_n \otimes H_n)$, defined by

$$W = \sum_{r=1}^{n}\sum_{s=1}^{n} |\,\boldsymbol{y}_r\rangle\langle \boldsymbol{y}_s\,| \otimes |\,\boldsymbol{x}_r\rangle\langle \boldsymbol{x}_s\,|$$
$$= \sum_{r=1}^{n}\sum_{s=1}^{n} |\,\boldsymbol{y}_r \otimes \boldsymbol{x}_r\rangle\langle \boldsymbol{y}_s \otimes \boldsymbol{x}_s\,|$$
$$= |\sum_{r=1}^{n} \boldsymbol{y}_r \otimes \boldsymbol{x}_r\rangle\langle \sum_{s=1}^{n} \boldsymbol{y}_s \otimes \boldsymbol{x}_s\,|,$$

is clearly positive, since it can be directly read from the last representation that W is a constant multiple of a one-dimensional projection. Since V is completely positive, it follows that

$$(I_n \otimes V)(W) = \sum_{r=1}^{n}\sum_{s=1}^{n} |\,\boldsymbol{y}_r\rangle\langle \boldsymbol{y}_s\,| \otimes V(|\,\boldsymbol{x}_r\rangle\langle \boldsymbol{x}_s\,|) \tag{8.57}$$

is a positive mapping. By Lemma (8.4.5), there exists an orthonormal basis $\{\boldsymbol{v}_1, \ldots, \boldsymbol{v}_{n^2}\}$ of $H_n \otimes H_n$ and positive numbers $\lambda_1, \ldots, \lambda_{n^2}$ such that

$$(I_n \otimes V)(W) = \sum_{r=1}^{n}\sum_{s=1}^{n} |\,\boldsymbol{y}_r\rangle\langle \boldsymbol{y}_s\,| \otimes V(|\,\boldsymbol{x}_r\rangle\langle \boldsymbol{x}_s\,|) = \sum_{i=1}^{n^2} \lambda_i\,|\,\boldsymbol{v}_i\rangle\langle \boldsymbol{v}_i\,|,$$

which can also be written as

$$(I_n \otimes V)(W) = \sum_{i=1}^{n^2} |\,\sqrt{\lambda_i}\boldsymbol{v}_i\rangle\langle \sqrt{\lambda_i}\boldsymbol{v}_i\,|\,. \tag{8.58}$$

If

$$\boldsymbol{v} = \sum_{j=1}^{n}\sum_{i=1}^{n} v_{ij}\boldsymbol{y}_j \otimes \boldsymbol{x}_i$$

is a vector in $H_n \otimes H_n$, we associate with $\boldsymbol{v}$ a mapping $V_{\boldsymbol{v}} \in L(H_n)$ by

$$V_{\boldsymbol{v}} = \sum_{i=1}^{n}\sum_{j=1}^{n} v_{ij}\,|\,\boldsymbol{x}_i\rangle\langle \boldsymbol{x}_j\,|\,. \tag{8.59}$$

Then, a straightforward computation gives

$$|\,\boldsymbol{v}\rangle\langle \boldsymbol{v}\,| = \sum_{r=1}^{n}\sum_{s=1}^{n} |\,\boldsymbol{y}_r\rangle\langle \boldsymbol{y}_s\,| \otimes V_{\boldsymbol{v}}\,|\,\boldsymbol{x}_r\rangle\langle \boldsymbol{x}_s\,|\,V_{\boldsymbol{v}}^{\,*}. \tag{8.60}$$

Associating to each $\sqrt{\lambda_i}\boldsymbol{v}_i$ in (8.58), mapping V_i as in (8.59), and using (8.60) we see that

$$(I_n \otimes V)(W) = \sum_{i=1}^{n^2} \sum_{r=1}^{n} \sum_{s=1}^{n} |\boldsymbol{y}_r\rangle\langle\boldsymbol{y}_s| \otimes V_i\, |\boldsymbol{x}_r\rangle\langle\boldsymbol{x}_s|\, V_i^*$$
$$= \sum_{r=1}^{n} \sum_{s=1}^{n} |\boldsymbol{y}_r\rangle\langle\boldsymbol{y}_s| \otimes \sum_{i=1}^{n^2} V_i\, |\boldsymbol{x}_r\rangle\langle\boldsymbol{x}_s|\, V_i^*,$$

which, together with (8.57), gives for each pair r, s,

$$V(|\boldsymbol{x}_r\rangle\langle\boldsymbol{x}_s|) = \sum_{i=1}^{n^2} V_i\, |\boldsymbol{x}_r\rangle\langle\boldsymbol{x}_s|\, V_i^*.$$

Since mappings $|\boldsymbol{x}_r\rangle\langle\boldsymbol{x}_s|$ span the whole space $L(H_n)$, equation

$$V(A) = \sum_{i=1}^{n^2} V_i A V_i^*$$

holds for each $A \in L(H_n)$. □

Another quite useful criterion for complete positiveness can be gathered from the proof of the previous theorem:

Theorem 8.4.6. *A linear mapping $V : L(H_n) \to L(H_n)$ is completely positive if and only if*

$$T = \sum_{r=1}^{n} \sum_{s=1}^{n} |\boldsymbol{y}_r\rangle\langle\boldsymbol{y}_s| \otimes V(|\boldsymbol{x}_r\rangle\langle\boldsymbol{x}_s|)$$

is a positive mapping $H_n \otimes H_n \to H_n \otimes H_n$.

Proof. If $V \in L(H_n)$ is a completely positive mapping, then T is positive by the very definition, since

$$\sum_{r=1}^{n} \sum_{s=1}^{n} |\boldsymbol{y}_r\rangle\langle\boldsymbol{y}_s| \otimes |\boldsymbol{x}_r\rangle\langle\boldsymbol{x}_s|$$

is positive. On the other hand, if T is positive, then a representation

$$V(A) = \sum_{i=1}^{n^2} V_i A V_i^*$$

can be found as in the proof of the previous theorem. □

Recall that we also required that V should preserve the trace.

Lemma 8.4.6. *Let $V : L(H_n) \to L(H_n)$ be a completely positive mapping represented as $V(A) = \sum_{i=1}^{n^2} V_i A V_i^*$. Then V preserves the trace if and only if $\sum_{i=1}^{n^2} V_i^* V_i = I$.*

Proof. Let $\{\boldsymbol{x}_1, \ldots, \boldsymbol{x}_n\}$ be an orthonormal basis of H_n. Since

$$I = \sum_{r=1}^{n} |\,\boldsymbol{x}_r\rangle\langle \boldsymbol{x}_r\,|,$$

by direct calculation we find that

$$\begin{aligned}
&\langle \boldsymbol{x}_k \mid \sum_{i=1}^{n^2} V_i^* V_i \boldsymbol{x}_l\rangle \\
&= \sum_{i=1}^{n^2} \langle \boldsymbol{x}_k \mid V_i^* \sum_{r=1}^{n} |\,\boldsymbol{x}_r\rangle\langle \boldsymbol{x}_r \mid V_i \boldsymbol{x}_l\rangle \\
&= \sum_{i=1}^{n^2}\sum_{r=1}^{n} \langle \boldsymbol{x}_r \mid V_i \boldsymbol{x}_l\rangle\langle \boldsymbol{x}_k \mid V_i^* \boldsymbol{x}_r\rangle \\
&= \sum_{i=1}^{n^2}\sum_{r=1}^{n} \langle \boldsymbol{x}_r \mid V_i \,|\boldsymbol{x}_l\rangle\langle \boldsymbol{x}_k|\, V_i^* \boldsymbol{x}_r\rangle \\
&= \mathrm{Tr}\Big(\sum_{i=1}^{n^2} V_i \,|\boldsymbol{x}_l\rangle\langle \boldsymbol{x}_k|\, V_i^*\Big) = \mathrm{Tr}\big(V(|\,\boldsymbol{x}_l\rangle\langle \boldsymbol{x}_k\,|)\big).
\end{aligned}$$

If $\sum_{i=1}^{n^2} V_i^* V_i = I$, then

$$\mathrm{Tr}\big(V(|\,\boldsymbol{x}_l\rangle\langle \boldsymbol{x}_k\,|)\big) = \langle \boldsymbol{x}_k \mid \boldsymbol{x}_l\rangle = \delta_{kl} = \mathrm{Tr}(|\,\boldsymbol{x}_l\rangle\langle \boldsymbol{x}_k\,|),$$

i.e., V preserves the trace of the mappings $|\,\boldsymbol{x}_l\rangle\langle \boldsymbol{x}_k\,|$. Since V and the trace are linear and the mappings $|\,\boldsymbol{x}_l\rangle\langle \boldsymbol{x}_k\,|$ generate the whole space $L(H_n)$, V preserves all traces. On the other hand, if V preserves all traces, then $\sum_{i=1}^{n^2} V_i^* V_i = I$, since any linear mapping $A \in L(H_n)$ is determined by the values $\langle \boldsymbol{x}_k \mid A\boldsymbol{x}_l\rangle$. □

The second structure theorem for completely positive mappings connects them to unitary mappings and offers an interesting interpretation.

Theorem 8.4.7. *A linear mapping $V : L(H_n) \to L(H_n)$ is completely positive and trace preserving if and only if there exists a unitary mapping $U \in L(H_n \otimes H_{n^2})$ and a pure state $B = |\,\boldsymbol{b}\rangle\langle \boldsymbol{b}\,|$ of system $L(H_{n^2})$ such that*

$$V(A) = \mathrm{Tr}_{H_{n^2}}(U(A \otimes B)U^*)$$

for each $A \in L(H_n)$.

Remark 8.4.7. A unitary time evolution

$$|\,\boldsymbol{x}\rangle\langle \boldsymbol{x}\,| \mapsto |\,U\boldsymbol{x}\rangle\langle U\boldsymbol{x}\,|$$

can be rewritten as

$$|\,\boldsymbol{x}\rangle\langle \boldsymbol{x}\,| \mapsto U\,|\,\boldsymbol{x}\rangle\langle \boldsymbol{x}\,|\,U^*,$$

so the above theorem states that any completely positive mapping $L(H_n) \to L(H_n)$ can be interpreted as a unitary time evolution in a larger system.

Proof. Assume first that $V(A) = \mathrm{Tr}_{H_{n^2}}(U(A \otimes B)U^*)$. Let $\{\boldsymbol{x}_1, \ldots, \boldsymbol{x}_n\}$ and $\{\boldsymbol{y}_1, \ldots, \boldsymbol{y}_{n^2}\}$ be orthonormal bases of H_n and H_{n^2}, respectively. We write U^* in the form

$$U^* = \sum_{r=1}^{n}\sum_{s=1}^{n} |\,\boldsymbol{x}_r\rangle\langle \boldsymbol{x}_s\,| \otimes U^*_{rs}$$

and define $V_k \in L(H_n)$ as

$$V_k = \sum_{i=1}^{n}\sum_{j=1}^{n} \langle U^*_{ji}\boldsymbol{y}_k \mid \boldsymbol{b}\rangle\, |\,\boldsymbol{x}_i\rangle\langle \boldsymbol{x}_j\,|\,.$$

Now, if

$$A = \sum_{i=1}^{n}\sum_{j=1}^{n} a_{ij}\, |\,\boldsymbol{x}_i\rangle\langle \boldsymbol{x}_j\,|,$$

a direct calculation gives

$$V_k A V_k{}^* = \sum_{i=1}^{n}\sum_{j=1}^{n}\sum_{r=1}^{n}\sum_{t=1}^{n} \langle \boldsymbol{x}_r \mid A\boldsymbol{x}_t\rangle\langle U^*_{ri}\boldsymbol{y}_k \,|\,|\,\boldsymbol{b}\rangle\langle \boldsymbol{b}\,|\, U^*_{tj}\boldsymbol{y}_k\rangle\, |\,\boldsymbol{x}_i\rangle\langle \boldsymbol{x}_j\,|\,. \quad (8.61)$$

On the other hand, (8.45) gives that

$$\begin{aligned}
&\mathrm{Tr}_{H_{n^2}}(U(A \otimes B)U^*)\\
&= \sum_{i=1}^{n}\sum_{j=1}^{n}\sum_{k=1}^{n^2} \langle \boldsymbol{x}_i \otimes \boldsymbol{y}_k \mid U(A \otimes B)U^*(\boldsymbol{x}_j \otimes \boldsymbol{y}_k)\rangle\, |\,\boldsymbol{x}_i\rangle\langle \boldsymbol{x}_j\,|\\
&= \sum_{k=1}^{n^2}\sum_{i=1}^{n}\sum_{j=1}^{n}\sum_{r=1}^{n}\sum_{t=1}^{n} \langle \boldsymbol{x}_r \otimes U^*_{ri}\boldsymbol{y}_k \mid A\boldsymbol{x}_t \otimes BU^*_{tj}\boldsymbol{y}_k\rangle\, |\,\boldsymbol{x}_i\rangle\langle \boldsymbol{x}_j\,|\\
&= \sum_{k=1}^{n^2}\sum_{i=1}^{n}\sum_{j=1}^{n}\sum_{r=1}^{n}\sum_{t=1}^{n} \langle \boldsymbol{x}_r \mid A\boldsymbol{x}_t\rangle\langle U^*_{ri}\boldsymbol{y}_k \,|\,|\,\boldsymbol{b}\rangle\langle \boldsymbol{b}\,|\, U^*_{tj}\boldsymbol{y}_k\rangle\, |\,\boldsymbol{x}_i\rangle\langle \boldsymbol{x}_j\,|,
\end{aligned}$$

which, using (8.61), can be written as

$$\mathrm{Tr}_{H_{n^2}}(U(A \otimes B)U^*) = \sum_{k=1}^{n^2} V_k A V_k{}^*.$$

Thus, by Theorem (8.4.5), V is completely positive. To see that V is trace preserving, notice that a unitary mapping U converts an orthonormal basis B to another orthonormal basis B', so

$$\begin{aligned}\mathrm{Tr}(UTU^*) &= \sum_{\boldsymbol{z}'\in B} \langle \boldsymbol{z}' \mid UTU^*\boldsymbol{z}'\rangle = \sum_{\boldsymbol{z}'\in B'} \langle U^*\boldsymbol{z}' \mid TU^*\boldsymbol{z}'\rangle \\ &= \sum_{\boldsymbol{z}'\in B'} \langle U^{-1}\boldsymbol{z}' \mid TU^{-1}\boldsymbol{z}'\rangle = \sum_{\boldsymbol{z}\in B} \langle \boldsymbol{z} \mid T\boldsymbol{z}\rangle = \mathrm{Tr}(T).\end{aligned}$$

Since "tracing over" $\mathrm{Tr}_{H_{n^2}} : L(H_n \otimes H_{n^2}) \to L(H_n)$, $T \mapsto T_1 = \mathrm{Tr}_{H_{n^2}}(T)$ is defined by condition

$$\mathrm{Tr}(T(P \otimes I_{n^2})) = \mathrm{Tr}(T_1 P)$$

for each projection $P \in L(H_n)$, we can use $P = I_n$ to obtain

$$\begin{aligned}&\mathrm{Tr}(\mathrm{Tr}_{H_{n^2}}(U(A \otimes B)U^*)) \\ &= \mathrm{Tr}(U(A \otimes B)U^*) = \mathrm{Tr}(A \otimes B) = \mathrm{Tr}(A)\mathrm{Tr}(B) = \mathrm{Tr}(A).\end{aligned}$$

On the other hand, let $V : L(H_n) \to L(H_n)$ be a completely positive and trace-preserving linear mapping. By Theorem 8.4.5, V can be represented as

$$V(A) = \sum_{i=1}^{n^2} V_i A V_i^*,$$

where each $V_i \in L(H_n)$ and

$$\sum_{i=1}^{n^2} V_i V_i^* = I.$$

Fix an arbitrary vector $\boldsymbol{b} \in H_{n^2}$ of unit length. For any $\boldsymbol{x} \in H_n$, we define

$$U(\boldsymbol{x} \otimes \boldsymbol{b}) = \sum_{i=1}^{n^2} V_i \boldsymbol{x} \otimes \boldsymbol{y}_i. \tag{8.62}$$

Clearly U is linear with respect to $\boldsymbol{x}$. If $\boldsymbol{x}_1, \boldsymbol{x}_2 \in H_n$, a direct computation gives

$$\begin{aligned}\langle U(\boldsymbol{x}_1 \otimes \boldsymbol{b}) \mid U(\boldsymbol{x}_2 \otimes \boldsymbol{b})\rangle &= \sum_{i=1}^{n^2}\sum_{j=1}^{n^2} \langle V_i\boldsymbol{x}_1 \otimes \boldsymbol{y}_i \mid V_j\boldsymbol{x}_2 \otimes \boldsymbol{y}_j\rangle \\ &= \sum_{i=1}^{n^2}\sum_{j=1}^{n^2} \langle V_i\boldsymbol{x}_1 \mid V_j\boldsymbol{x}_2\rangle\langle \boldsymbol{y}_i \mid \boldsymbol{y}_j\rangle = \sum_{i=1}^{n^2} \langle V_i\boldsymbol{x}_1 \mid V_i\boldsymbol{x}_2\rangle \\ &= \sum_{i=1}^{n^2} \langle \boldsymbol{x}_1 \mid V_i^* V_i \boldsymbol{x}_2\rangle = \langle \boldsymbol{x}_1 \mid \boldsymbol{x}_2\rangle \\ &= \langle \boldsymbol{x}_1 \otimes \boldsymbol{b} \mid \boldsymbol{x}_2 \otimes \boldsymbol{b}\rangle.\end{aligned}$$

Denoting

$$H_n \otimes \boldsymbol{b} = \{\boldsymbol{x} \otimes \boldsymbol{b} \mid \boldsymbol{x} \in H_n\} \subseteq H_n \otimes H_{n^2},$$

we see that $U : H_n \otimes \boldsymbol{b} \to H_n \otimes H_{n^2}$ is a linear mapping that preserves inner products. It follows that U can be extended (usually, in many ways) to a unitary mapping $U' \in L(H_n \otimes H_{n^2})$. We fix one extension and denote that again by U. If $\boldsymbol{x}$ is any unit-length vector, then,

$$\begin{aligned} U(|\boldsymbol{x}\rangle\langle\boldsymbol{x}| \otimes |\boldsymbol{b}\rangle\langle\boldsymbol{b}|)U^* &= U\ |\boldsymbol{x}\otimes\boldsymbol{b}\rangle\langle\boldsymbol{x}\otimes\boldsymbol{b}|\ U^* \\ &= |U(\boldsymbol{x}\otimes\boldsymbol{b})\rangle\langle U(\boldsymbol{x}\otimes\boldsymbol{b})| \\ &= |\sum_{i=1}^{n^2} V_i\boldsymbol{x}\otimes\boldsymbol{y}_i\rangle\langle\sum_{i=1}^{n^2} V_i\boldsymbol{x}\otimes\boldsymbol{y}_i|. \end{aligned}$$

Using Example 8.4.4, it is easy to see that

$$\begin{aligned} \mathrm{Tr}_{H_{n^2}}(U(|\boldsymbol{x}\rangle\langle\boldsymbol{x}| \otimes |\boldsymbol{b}\rangle\langle\boldsymbol{b}|)U^*) &= \sum_{i=1}^{n^2} |V_i\boldsymbol{x}\rangle\langle V_i\boldsymbol{x}| \\ &= \sum_{i=1}^{n^2} V_i\ |\boldsymbol{x}\rangle\langle\boldsymbol{x}|\ V_i^* \\ &= V(|\boldsymbol{x}\rangle\langle\boldsymbol{x}|). \end{aligned}$$

Hence with U defined as in (8.62) and $B = |\boldsymbol{b}\rangle\langle\boldsymbol{b}|$, the claim holds at least for all pure states $|\boldsymbol{x}\rangle\langle\boldsymbol{x}|$. Since linear mappings

$$A \mapsto \mathrm{Tr}_{H_{n^2}}(U(A \otimes B)U^*)$$

and $A \mapsto V(A)$ agree on all pure states and all states can be expressed as convex combinations of pure states, the mappings must be equal. □

8.4.6 Jozsa's Theorem of Cloning and Deleting

Recently, Richard Jozsa introduced a stronger variant of the no-cloning principle [48]. Jozsa's theorem can also be extended to cover the no-deleting principle by Pati and Braunstein [66].

Theorem 8.4.8 (Jozsa's No-Cloning Theorem). *Let $\boldsymbol{x}_1, \ldots, \boldsymbol{x}_k$ be unit vectors in H_n such that $\langle\boldsymbol{x}_i \mid \boldsymbol{x}_j\rangle \neq 0$ for each pair i, j. Let $\boldsymbol{0} \in H_n$ also be a unit vector of H_n and $\rho_1, \ldots, \rho_k$ some states in $L(H_m)$. There exists a completely positive mapping $V \in (H_n \otimes H_n \otimes H_m)$ such that*

$$V(\ |\boldsymbol{x}_i\rangle\langle\boldsymbol{x}_i| \otimes |\boldsymbol{0}\rangle\langle\boldsymbol{0}| \otimes\rho_i) = |\boldsymbol{x}_i\rangle\langle\boldsymbol{x}_i| \otimes |\boldsymbol{x}_i\rangle\langle\boldsymbol{x}_i| \otimes\rho_i'$$

for each $i \in \{1, \ldots, k\}$ if and only if there exists a completely positive mapping $V' \in L(H_n \otimes H_m)$ such that

$$V'(\ |\boldsymbol{0}\rangle\langle\boldsymbol{0}| \otimes\rho_i) = |\boldsymbol{x}_i\rangle\langle\boldsymbol{x}_i| \otimes\rho_i''$$

for each $i \in \{1, \ldots, k\}$.

Remark 8.4.8. The above theorem can be interpreted as follows: There exists a physical operation that produces copies of pure states $|\boldsymbol{x}_k\rangle$, ..., $|\boldsymbol{x}_k\rangle$ (with extra information ρ_1, ..., ρ_k such that ρ_i belongs to state $|\boldsymbol{x}_i\rangle$) if and only if there exists a physical operation that produces state $|\boldsymbol{x}_i\rangle$ from ancillary information ρ_i. Choosing $\rho_1 = \ldots = \rho_k = \rho$, we get the classical no-cloning theorem of Wootters and Zurek (Theorem 2.3.1).

Proof. If mapping $V' \in L(H_n \otimes H_m)$ satisfying

$$V'(\,|\mathbf{0}\rangle\langle\mathbf{0}|\otimes\rho_i) = |\boldsymbol{x}_i\rangle\langle\boldsymbol{x}_i|\otimes\rho_i''$$

exists, then the desired mapping V could be easily obtained as an extension.

Assume then that mapping V as stated in the theorem exists. By Lemma 8.4.3, each state $\rho_i \in H_m$ can be represented as a partial trace over a pure state determined by some vector $\boldsymbol{y}_i \in H_m \otimes H_m$. Moreover, by Theorem 8.4.7, there exists an environment H_e such that V can be interpreted as a unitary mapping in space $H_n \otimes H_n \otimes H_m \otimes H_m \otimes H_e$:

$$U(\boldsymbol{x}_i \otimes \mathbf{0} \otimes \boldsymbol{y}_i \otimes \boldsymbol{e}) = \boldsymbol{x}_i \otimes \boldsymbol{x}_i \otimes \boldsymbol{z}_i,$$

where $\boldsymbol{e} \in H_e$ and $\boldsymbol{z}_i \in H_m \otimes H_m \otimes H_e$. Now that U is unitary, we have by Lemma 8.2.5, that

$$\langle \boldsymbol{x}_i \otimes \mathbf{0} \otimes \boldsymbol{y}_i \otimes \boldsymbol{e} \mid \boldsymbol{x}_j \otimes \mathbf{0} \otimes \boldsymbol{y}_j \otimes \boldsymbol{e}\rangle = \langle \boldsymbol{x}_i \otimes \boldsymbol{x}_i \otimes \boldsymbol{z}_i \mid \boldsymbol{x}_j \otimes \boldsymbol{x}_j \otimes \boldsymbol{z}_j\rangle \quad (8.63)$$

for each i, $j \in \{1, \ldots, k\}$. Equation 8.63 can be written as

$$\langle \boldsymbol{x}_i \mid \boldsymbol{x}_j\rangle\langle \mathbf{0} \otimes \boldsymbol{y}_i \otimes \boldsymbol{e} \mid \mathbf{0} \otimes \boldsymbol{y}_j \otimes \boldsymbol{e}\rangle = \langle \boldsymbol{x}_i \mid \boldsymbol{x}_j\rangle\langle \boldsymbol{x}_i \otimes \boldsymbol{z}_i \mid \boldsymbol{x}_j \otimes \boldsymbol{z}_j\rangle. \quad (8.64)$$

By the assumption, $\langle \boldsymbol{x}_i \mid \boldsymbol{x}_j\rangle \neq 0$ always, and (8.64) yields

$$\langle \mathbf{0} \otimes \boldsymbol{y}_i \otimes \boldsymbol{e} \mid \mathbf{0} \otimes \boldsymbol{y}_j \otimes \boldsymbol{e}\rangle = \langle \boldsymbol{x}_i \otimes \boldsymbol{z}_i \mid \boldsymbol{x}_j \otimes \boldsymbol{z}_j\rangle \quad (8.65)$$

for each $i, j \in \{1, \ldots, k\}$. Now, by Lemma 8.2.8, we know that there is a unitary mapping $U' \in L(H_n \otimes H_m \otimes H_m \otimes H_e)$ such that

$$U'(\mathbf{0} \otimes \boldsymbol{y}_i \otimes \boldsymbol{e}) = \boldsymbol{x}_i \otimes \boldsymbol{z}_i$$

for each $i \in \{1, \ldots, k\}$. By tracing over H_e and the other copy of H_m, we get the desired mapping V'. □

The no-deleting principle by Pati and Braunstein [66] is the following: If $\boldsymbol{x}_1$, ..., $\boldsymbol{x}_k$ contains no orthogonal states, then no physical operation can delete another copy of $\boldsymbol{x}_i$. That is, operation

$$|\boldsymbol{x}_i\rangle\,|\boldsymbol{x}_i\rangle \mapsto |\boldsymbol{x}\rangle\,|\mathbf{0}\rangle$$

is impossible in a sense that will be clarified in the statement of the next theorem.

In fact, the deletion operation is not impossible: in [66], it is emphasized that we can, for instance, include some environment and just define operation

$$|\boldsymbol{x}_i\rangle\,|\boldsymbol{x}_i\rangle\,|\boldsymbol{e}\rangle \mapsto |\boldsymbol{x}_i\rangle\,|\boldsymbol{0}\rangle\,|\boldsymbol{e}_i\rangle$$

which is essentially only a swap between the environment state and the second copy of the state $\boldsymbol{x}_i$. In this case, the state $\boldsymbol{x}_i$ remains in the environment, and it can be recovered perfectly.

On the other hand, if the projection postulate is adopted, then deletion is possible, as pointed out in [48]. For deletion, then, it is enough to observe the second copy of state $|\boldsymbol{x}_i\rangle\,|\boldsymbol{x}_i\rangle$ and then to swap the resulting state into $|\boldsymbol{0}\rangle$.

Theorem 8.4.9 (No-Deleting Principle). *Assume that* $\boldsymbol{x}_1, \ldots, \boldsymbol{x}_k \in H_n$ *and* $\boldsymbol{0} \in H_n$ *are as above. If there exists a completely positive mapping erasing the second copy of* $\boldsymbol{x}_i$, *i.e., a mapping for which*

$$|\,\boldsymbol{x}_i\rangle\langle\boldsymbol{x}_i\,| \otimes |\,\boldsymbol{x}_i\rangle\langle\boldsymbol{x}_i\,| \mapsto |\,\boldsymbol{x}_i\rangle\langle\boldsymbol{x}_i\,| \otimes |\,\boldsymbol{0}\rangle\langle\boldsymbol{0}\,|$$

holds, then vectors $\boldsymbol{x}_i$ *can be recovered perfectly from the environment.*

Proof. Assume that there is a completely positive mapping for which the assumption holds. By Theorem 8.4.7, there is a space H_e, vectors $\boldsymbol{e}$, $\boldsymbol{e}_1, \ldots, \boldsymbol{e}_k \in H_n$, and a unitary $U \in L(H_n \otimes H_n \otimes H_e)$ that satisfies

$$U(\boldsymbol{x}_i \otimes \boldsymbol{x}_i \otimes \boldsymbol{e}) = \boldsymbol{x}_i \otimes \boldsymbol{0} \otimes \boldsymbol{e}_i.$$

By Lemma 8.2.5, for each $i, j \in \{1, \ldots, k\}$,

$$\langle \boldsymbol{x}_i \otimes \boldsymbol{x}_i \otimes \boldsymbol{e} \mid \boldsymbol{x}_j \otimes \boldsymbol{x}_j \otimes \boldsymbol{e}\rangle = \langle \boldsymbol{x}_i \otimes \boldsymbol{0} \otimes \boldsymbol{e}_i \mid \boldsymbol{x}_j \otimes \boldsymbol{0} \otimes \boldsymbol{e}_j\rangle,$$

which can be written as

$$\langle \boldsymbol{x}_i \mid \boldsymbol{x}_j\rangle\langle \boldsymbol{x}_i \otimes \boldsymbol{e} \mid \boldsymbol{x}_j \otimes \boldsymbol{e}\rangle = \langle \boldsymbol{x}_i \mid \boldsymbol{x}_j\rangle\langle \boldsymbol{0} \otimes \boldsymbol{e}_i \mid \boldsymbol{0} \otimes \boldsymbol{e}_j\rangle,$$

or as

$$\langle \boldsymbol{x}_i \otimes \boldsymbol{e} \mid \boldsymbol{x}_j \otimes \boldsymbol{e}\rangle = \langle \boldsymbol{0} \otimes \boldsymbol{e}_i \mid \boldsymbol{0} \otimes \boldsymbol{e}_j\rangle. \tag{8.66}$$

Lemma 8.2.8 states that there is a unitary mapping $U' \in L(H_n \otimes H_e)$ such that

$$U'(\boldsymbol{x}_i \otimes \boldsymbol{e}) = \boldsymbol{0} \otimes \boldsymbol{e}_i$$

for each $i \in \{1, \ldots, k\}$. As a unitary mapping, U' also has an inverse, and this implies that $\boldsymbol{x}_i$ can be recovered from the environment. □

8.5 Exercises

1. Show that the trace function

$$\mathrm{Tr}(A) = \sum_{i=1}^{n} \langle \boldsymbol{x}_i \mid A\boldsymbol{x}_i \rangle$$

is independent of the chosen orthonormal basis $\{\boldsymbol{x}_1, \ldots, \boldsymbol{x}_n\}$.
2. Show that if P is a self-adjoint operator $H_n \to H_n$ and $P^2 = P$, then there is a subspace $W \in H_n$ such that $P = P_W$.
3. If $\lambda_1, \ldots, \lambda_n$ are eigenvalues of a self-adjoint operator A and $\boldsymbol{x}_1, \ldots, \boldsymbol{x}_n$ the corresponding orthonormal set of eigenvectors, verify that

$$A = \lambda_1 \, |\boldsymbol{x}_1\rangle\langle \boldsymbol{x}_1| + \ldots + \lambda_n \, |\boldsymbol{x}_1\rangle\langle \boldsymbol{x}_n| \, .$$

4. Verify that the polarization equation

$$\langle \boldsymbol{x} \mid \boldsymbol{y} \rangle = \frac{1}{4} \sum_{k=0}^{3} i^k \langle \boldsymbol{y} + i^k \boldsymbol{x} \mid \boldsymbol{y} + i^k \boldsymbol{x} \rangle$$

holds. Conclude that if $||A\boldsymbol{x}|| = ||\boldsymbol{x}||$ for any $\boldsymbol{x} \in H_n$, then also $\langle A\boldsymbol{x} \mid A\boldsymbol{y} \rangle = \langle \boldsymbol{x} \mid \boldsymbol{y} \rangle$ for each $\boldsymbol{x}$, $\boldsymbol{y} \in H_n$.
5. Prove that

$$\lambda(|\boldsymbol{x}_1\rangle\langle \boldsymbol{x}_1| + |\boldsymbol{x}_2\rangle\langle \boldsymbol{x}_2|) = \lambda(|\boldsymbol{x}_1'\rangle\langle \boldsymbol{x}_1'| + |\boldsymbol{x}_2'\rangle\langle \boldsymbol{x}_2'|),$$

where

$$\boldsymbol{x}_1' = \boldsymbol{x}_1 \cos\alpha - \boldsymbol{x}_2 \sin\alpha,$$
$$\boldsymbol{x}_2' = \boldsymbol{x}_1 \sin\alpha + \boldsymbol{x}_2 \cos\alpha,$$

and α is any real number.
6. Prove that $| A\boldsymbol{x}\rangle\langle B\boldsymbol{y} |= A \mid \boldsymbol{x}\rangle\langle \boldsymbol{y} \mid B^*$ for each $\boldsymbol{x}$, $\boldsymbol{y} \in H_n$ and any operators A, $B : H_n \to H_n$.
7. Derive the generalized Schrödinger equation

$$i\frac{d}{dt}\rho(t) = [H, \rho(t)]$$

with equation

$$\rho(t) = U(t)\rho(0)U(t)^*$$

and representation $U(t) = e^{-itH}$ as starting points.

9. Appendix B: Mathematical Background

The purpose of this chapter is to introduce the reader to the basic mathematical notions used in this book.

9.1 Group Theory

9.1.1 Preliminaries

A *group* G is a set equipped with mapping $G \times G \to G$, i.e., with a rule that unambiguously describes how to create one element of G out of an ordered pair of given ones. This operation is frequently called the *multiplication* or the *addition* and is denoted by $g = g_1 g_2$ or $g = g_1 + g_2$ respectively. Moreover, there is a special required element in G, which is called the *unit element* or the *neutral element* and it is usually denoted by 1 in the multiplicative notation, and 0 in the additive notation.

Furthermore, there is one more required operation, *inversion*, that sends any element of g into its *inverse element* g^{-1} (resp. *opposite element* $-g$ when additive notations are used). Finally, the group operations are required to obey the following *group axioms* (using multiplicative notations):

1. For all elements g_1, g_2, and g_3, $g_1(g_2 g_3) = (g_1 g_2) g_3$.
2. For any element g, $g1 = 1g = g$.
3. For any element g, $gg^{-1} = g^{-1}g = 1$.

If a group G also satisfies

4. For all elements g_1 and g_2, $g_1 g_2 = g_2 g_1$,

then G is called an *abelian* group or a *commutative* group.

To be precise, instead of speaking about group G, we should say that $(G, \cdot, ^{-1}, 1)$ is a group. Here G is a set of group elements; $\cdot$ stands for the multiplication; $^{-1}$ stands for the inversion, and 1 is the unit element. However, if there is no danger of confusion, we will just use the notation G for the group, as well as for the underlying set.

Example 9.1.1. Integers $\mathbb{Z}$ form an abelian group having addition as the group operation, 0 as the neutral element and mapping $n \mapsto -n$ as the inversion. On the other hand, natural numbers

$$\mathbb{N} = \{1, 2, 3, \ldots\}$$

do not form a group with respect to these operations, since $\mathbb{N}$ is not closed under the inversion $n \mapsto -n$, and there is no neutral element in $\mathbb{N}$.

Example 9.1.2. Nonzero complex numbers form an abelian group with respect to multiplication. The neutral element is 1, and the inversion is the mapping $z \mapsto z^{-1}$.

Example 9.1.3. The set of $n \times n$ matrices over $\mathbb{C}$ that have a nonzero determinant constitute a group with respect to matrix multiplication. This group, denoted by $GL_n(\mathbb{C})$ and called the *general linear group* over $\mathbb{C}$, is not abelian unless $n = 1$, when the group is essentially the same as in the previous example.

Regarding group axiom 1, it makes sense to omit the parenthesis and write just $g_1 g_2 g_3 = g_1(g_2 g_3) = (g_1 g_2) g_3$. This clearly generalizes to the products of more than three elements. A special case is a product $g \ldots g$ (k times), which we will denote by g^k (in the additive notations it would be kg). We also define $g^0 = 1$ and $g^{-k} = (g^{-1})^k$ ($0g = 0$, $-kg = k(-g)$ when written additively).

9.1.2 Subgroups, Cosets

A *subgroup* H of a group G is a group contained in G in the following way: H is contained in G as a set, and the group operations (product, inversion, and the unit element[1]) of H are only those of G which are restricted to H. If H is a subgroup of G, we write $H \leq G$.

Let us now suppose that a group G (now written multiplicatively) has H as a subgroup. Then, for each $g \in G$, the set

$$gH = \{gh \mid h \in H\} \tag{9.1}$$

is called a *coset* of H (determined by g).

Simple but useful observations can easily be made; the first is that each element $g \in G$ belongs to *some* coset, for instance in gH. This is true because H, as a subgroup, contains the neutral element 1 and therefore $g = g1 \in gH$. This observation tells us that the cosets of a subgroup H cover the whole group G. Notice that, especially for any $h \in H$, we have $hH = H$ because H, being a group, is closed under multiplication and each element $h_1 \in H$ appears in hH ($h_1 = h(h^{-1}h_1)$).

Other observations are given in the following lemmata.

Lemma 9.1.1. *$g_1 H = g_2 H$ if and only if $g_1^{-1} g_2 \in H$.*

[1] The unit element can be seen as a nullary operation.

Proof. If $g_1H = g_2H$, then by necessity $g_2 = g_1h$ for some $h \in H$. Hence, $g_1^{-1}g_2 = h \in H$. On the other hand, if $g_1^{-1}g_2 = h \in H$, then $g_2 = g_1h$ and so $g_2H = g_1hH = g_1H$. □

Remark 9.1.1. Notice that $g_1^{-1}g_2 \in H$ if and only if $g_2^{-1}g_1 \in H$. This is true because H contains all inverses of the elements in H and $(g_1^{-1}g_2)^{-1} = g_2^{-1}g_1$. In the additive notations, the condition $g_1^{-1}g_2 \in H$ would be written as $-g_1 + g_2 \in H$.

Lemma 9.1.2. *Let H be a finite subgroup of G. Then all the cosets of H have $m = |H|$ elements.*

Proof. By the very definition of (9.1), it is clear that each coset can have *at most* m elements. If some coset gH has less than m elements, then for some $h_i \neq h_j$ we must have $gh_i = gh_j$. Multiplication by g^{-1}, however, gives $h_i = h_j$, which is a contradiction. □

Definition 9.1.1. *If $g_1H = g_2H$, we say that g_1 is congruent to g_2 modulo H.*

Lemma 9.1.3. *If $g_1H \neq g_2H$, then also $g_1H \cap g_2H = \emptyset$.*

Proof. Assume, on the contrary, that $g_1H \neq g_2H$ have a common element g. Since $g \in g_1H \cap g_2H$, we must have $g = g_1h_1 = g_2h_2$ for some elements $h_1, h_2 \in H$. But then $g_1 = g_2h_2h_1^{-1}$, and $g_1H = g_2h_2h_1^{-1}H = g_2h_2H = g_2H$, a contradiction. □

If G is finite and $H \leq G$, we call the number of the cosets of H the *index* of H in G and denote this number by $[G : H]$. Since the cosets cover G, we know by Lemma 9.1.3 that the group G is a disjoint union of $[G : H]$ cosets of H, which all have the same cardinality $|H|$ by Lemma 9.1.2. Thus, we have obtained the following theorem

Theorem 9.1.1 (Lagrange). $|G| = [G : H] \cdot |H|$.

Even though Lagrange's theorem was easy to derive, it has very deep implication: the group structure, which does not look very demanding at first glance, is complicated enough to heavily restrict the cardinalities of the subgroups; the subgroup cardinality $|H|$ must always divide $|G|$. It follows that a group G with a prime number cardinality can only have the *trivial subgroups*, the one consisting of only the unit element and the group G itself.

9.1.3 Factor Groups

A subgroup $H \leq G$ is called *normal* if, for any $g \in G$ and $h \in H$, always $ghg^{-1} \in H$. Notice that all the subgroups of an abelian group are normal. For a normal subgroup $H \leq G$ we define the *product of cosets* g_1H and g_2H by

$$(g_1H)(g_2H) = (g_1g_2)H. \tag{9.2}$$

But, as far as we know, two distinct elements can define the same coset: $g_1H = g_2H$ may hold even if $g_1 \neq g_2$. Can the coset product ever not be well-defined, i.e., that the product would depend on the *representatives* g_1 and g_2 which are chosen? The answer is no:

Lemma 9.1.4. *If H is a normal subgroup, $g_1H = g_1'H$ and $g_2H = g_2'H$, then also $(g_1g_2)H = (g_1'g_2')H$.*

Proof. By assumption and Lemma 9.1.1, $g_1^{-1}g_1' = h_1 \in H$ and $g_2^{-1}g_2' = h_2 \in H$. But since H is normal, we have

$$\begin{aligned}(g_1g_2)^{-1}(g_1'g_2') &= g_2^{-1}g_1^{-1}g_1'g_2' = g_2^{-1}h_1g_2' \\ &= g_2^{-1}h_1g_2g_2^{-1}g_2' = g_2^{-1}h_1g_2h_2 \in H.\end{aligned}$$

The conclusion that $g_2^{-1}h_1g_2h_2 \in H$ is due to the fact that $g_2^{-1}h_1g_2$ is in H because H is normal. Therefore, by Lemma 9.1.1, $(g_1g_2)H = (g_1'g_2')H$. □

The coset product offers us a method of defining the *factor group* G/H:

Definition 9.1.2. *Let G be a group and $H \leq G$ a normal subgroup. The factor group G/H has the cosets of H as the group elements and the coset product as the group operation. The neutral element of G/H is $1H = H$ and the inverse of a coset gH is $g^{-1}H$. The factor group is also called the quotient group .*

Using the above terminology, $|G/H| = [G : H]$.

Example 9.1.4. Consider $\mathbb{Z}$, the additive group of integers and a fixed integer n. Then, all the integers divisible by n form a subgroup

$$n\mathbb{Z} = \{\ldots - 3n, -2n, -n, 0, n, 2n, 3n, \ldots\},$$

as is easily verified. Any coset of the subgroup $n\mathbb{Z}$ looks like

$$k + n\mathbb{Z} = \{\ldots k - 3n, k - 2n, k - n, k, k + n, k + 2n, k + 3n \ldots\}$$

and two integers k_1, k_2 are congruent modulo $n\mathbb{Z}$ if and only if $k_1 + n\mathbb{Z} = k_2 + n\mathbb{Z}$ which, by Lemma 9.1.1, holds if and only if $k_2 - k_1 \in n\mathbb{Z}$, i.e, $k_2 - k_1$ is divisible by n. The factor group $\mathbb{Z}/(n\mathbb{Z})$ is usually denoted by $\mathbb{Z}_n$.

Definition 9.1.3. *If integers k_1 and k_2 are congruent modulo $n\mathbb{Z}$, we also say that k_1 and k_2 are congruent modulo n and denote $k_1 \equiv k_2 \pmod{n}$.*

Let us pay attention to the subgroups of a finite group G of special type. Pick any element $g \in G$ and consider the set

$$\{g^0, g^1, g^2, \ldots\}. \tag{9.3}$$

The set (9.3) is contained in G and must, therefore, be finite. It follows that, for some $i > j$, equality $g^i = g^j$ holds, and multiplication by $(g^j)^{-1}$ gives $g^{i-j} = 1$. This supplies the motivation to the following definition.

Definition 9.1.4. *Let G be a finite group and let $g \in G$ be any element. The smallest positive integer k such that $g^k = 1$ is called the order of g and denoted by $k = \text{ord}(g)$.*

It follows that, if $k = \text{ord}(g)$, then $g^{l+km} = g^l g^{km} = g^l 1^m = g^l$ for any integer m. Moreover, it is easy to see that the set (9.3) is an abelian subgroup of G. In fact, the set (9.3) is clearly closed under multiplication and the inverse of an element g^l can be expressed as g^{-l+km}, where m is any integer. We say that set (9.3) is a *cyclic subgroup of G generated by g* and denote that by

$$\langle g \rangle = \{g^0, g^1, g^2, \ldots, g^{k-1}\}. \tag{9.4}$$

According to Lagrange's theorem, $k = \text{ord}(g) = |\langle g \rangle|$ always divides $|G|$. Thus, we have obtained

Corollary 9.1.1. *For each $g \in G$, we have $g^{|G|} = 1$.*

Proof. Since $|G|$ is divisible by $k = \text{ord}(g)$, we can write $|G| = k \cdot l$ for some integer l, and then $g^{|G|} = g^{k \cdot l} = 1^l = 1$. □

9.1.4 Group $\mathbb{Z}_n^*$

Consider again the group $\mathbb{Z}_n$. Although the group operation is the coset addition, it is also easy to see that the *coset product*

$$(k_1 + n\mathbb{Z})(k_2 + n\mathbb{Z}) = k_1 k_2 + n\mathbb{Z}$$

is a well-defined concept. In fact, suppose that $k_1 + n\mathbb{Z} = k_1' + n\mathbb{Z}$ and $k_2 + n\mathbb{Z} = k_2' + n\mathbb{Z}$. Then $k_1 - k_1'$ and $k_2 - k_2'$ are divisible by n, and also $k_1k_2 - k_1'k_2' = (k_1 - k_1')k_2 + k_1'(k_2 - k_2')$ is divisible by n. Therefore, $k_1k_2 + n\mathbb{Z} = k_1'k_2' + n\mathbb{Z}$.

Coset $1 + n\mathbb{Z}$ is clearly a neutral element with respect to multiplication, but the cosets $k + n\mathbb{Z}$ *do not* form a multiplicative group, the reason being that the cosets $k + n\mathbb{Z}$ such that $\gcd(k, n) > 1$ do not have any inverse. To see this, assume that $k + n\mathbb{Z}$ has an inverse $k' + n\mathbb{Z}$, i.e., $kk' + n\mathbb{Z} = 1 + n\mathbb{Z}$. This means then that $kk' - 1$ is divisible by n and, hence. also divisible by $\gcd(n, k)$. But then 1 would also be divisible by $\gcd(n, k)$, which is absurd because $\gcd(n, k) > 1$.

To remove the problem of non-invertible elements, we define $\mathbb{Z}_n^*$ to be the set of all the cosets $k + n\mathbb{Z}$ that have a multiplicative inverse; thus, $\mathbb{Z}_n^*$ becomes a multiplicative group.[2] Previously, we saw that all cosets $k + n\mathbb{Z}$ with $\gcd(k, n) > 1$ do not belong to $\mathbb{Z}_n^*$ and next we will demonstrate that

[2] To be more algebraic, the group $\mathbb{Z}_n^*$ is the *unit group* of *ring* $\mathbb{Z}_n$. The fact that $n\mathbb{Z}$ is an *ideal* of ring $\mathbb{Z}$ automatically implies that the coset product is a well-defined concept.

$\mathbb{Z}_n^*$ consists exactly of cosets $k+n\mathbb{Z}$ such that $\gcd(k,n)=1$ (Notice that this property is independent of the representative k chosen). For that purpose we have to find an inverse for each such coset, which will be an easy task after the following lemma.

Lemma 9.1.5 (Bezout's identity). *For any natural numbers x, y there are integers a and b such that $ax+by=\gcd(x,y)$.*

Proof. By induction on $M=\max\{x,y\}$. If $M=1$, then necessarily $x=y=1$ and $2\cdot 1-1\cdot 1=1=\gcd(1,1)$. For the general case we can assume, without loss of generality, that $M=x>y$. Because $\gcd(x-y,y)$ is a divisor of $x-y$ and y, it also divides x, so $\gcd(x-y,y)\leq\gcd(x,y)$. Similarly, $\gcd(x,y)\leq\gcd(x-y,y)$ and, hence, $\gcd(x-y,y)=\gcd(x,y)$. We apply the induction hypothesis to the pair $(x-y,y)$ to get numbers a' and b' such that

$$a'(x-y)+b'y=\gcd(x-y,y)=\gcd(x,y),$$

which gives the required numbers $a=a'$, $b=b'-a'$. □

If $\gcd(k,n)=1$, we can use the previous lemma to find the inverse of the coset $k+n\mathbb{Z}$: let a and b be the integers such that $ak+bn=1$. We claim that the coset $a+n\mathbb{Z}$ is the required multiplicative inverse. But this is easily verified: since $ak-1$ is divisible by n, we have $ak+n\mathbb{Z}=1+n\mathbb{Z}$ and therefore $(a+n\mathbb{Z})(k+n\mathbb{Z})=1+n\mathbb{Z}$.

How many elements are there in group $\mathbb{Z}_n^*$? Since *all* the cosets of $n\mathbb{Z}$ can be represented as $k+n\mathbb{Z}$, where k ranges over the set $\{0,\ldots,n-1\}$, we have to find out how many of those k values satisfy the extra condition $\gcd(k,n)=1$. The number of such values of k is denoted by $\varphi(n)=|\mathbb{Z}_n^*|$ and is called *Euler's φ-function.*

Let us first consider the case that $n=p^m$ is a prime power. Then, only the numbers $0p$, $1p$, $2p$, ..., $(p^{m-1}-1)p$ in the set $\{0,1,\ldots,p^m-1\}$ *do not* satisfy the condition $\gcd(k,p^m)=1$. Therefore, $\varphi(p^m)=p^m-p^{m-1}$. Especially $\varphi(p)=p-1$ for prime numbers.

Assume now that $n=n_1\cdots n_r$, where numbers n_i are pairwise coprime, i.e., $\gcd(n_i,n_j)=1$ whenever $i\neq j$. We will demonstrate that then $\varphi(n)=\varphi(n_1)\cdots\varphi(n_r)$ and, for that purpose, we present a well-known result:

Theorem 9.1.2 (Chinese Remainder Theorem). *Let $n=n_1\cdots n_r$ with $\gcd(n_i,n_j)=1$ when $i\neq j$. Then, for given cosets $k_i+n_i\mathbb{Z}$, $i\in\{1,\ldots,r\}$ there exists a unique coset $k+n\mathbb{Z}$ of $n\mathbb{Z}$ such that for each $i\in\{1,\ldots,r\}$ $k+n_i\mathbb{Z}=k_i+n_i\mathbb{Z}$.*

Proof. Let $m_i=n/n_i=n_1\cdots n_{i-1}n_{i+1}\cdots n_r$. Then, clearly $\gcd(m_i,n_i)=1$, and, according to Lemma 9.1.5, $a_im_i+b_in_i=1$ for some integers a_i and b_i. Let

$$k=a_1m_1k_1+\ldots+a_rm_rk_r.$$

Then

$$k - k_i = a_1 m_1 k_1 + \ldots + (a_i m_i - 1)k_i + \ldots + a_r m_r k_r$$

is divisible by n_i, since each m_j with $j \neq i$ is divisible as well, and $a_i m_i - 1$ is also divisible by n_i because $a_i m_i + b_i n_i = 1$. It follows that $k + n_i\mathbb{Z} = k_i + n_i\mathbb{Z}$, which proves the existence of such a coset $k + n\mathbb{Z}$. If also $k' + n_i\mathbb{Z} = k_i + n_i\mathbb{Z}$ for each i, then $k' - k$ is divisible by each n_i and, since the numbers n_i are coprime, $k' - k$ is also divisible by $n = n_1 \cdots n_r$. Hence, $k' + n\mathbb{Z} = k + n\mathbb{Z}$. □

Clearly, any coset $k + n\mathbb{Z} \in \mathbb{Z}_n^*$ defines r cosets $k + n_i\mathbb{Z} \in \mathbb{Z}_{n_i}^*$. But in accordance with Chinese Remainder Theorem, *all* the cosets $k_i + n_i\mathbb{Z} \in \mathbb{Z}_{n_i}^*$ are obtained in this way. To show that, we must demonstrate that, if we have $\gcd(k_i, n_i) = 1$ for each i, then the k which is given by the Chinese Remainder Theorem also satisfies $\gcd(k, n) = 1$. But this is straightforward: if $\gcd(k, n) > 1$, then also $d = \gcd(k', n_i) > 1$ for some i and some k' such that $k = k'k''$. Then, however, $k'k'' + n_i\mathbb{Z} = k + n_i\mathbb{Z} \notin \mathbb{Z}_n^*$. It follows that

$$\left|\mathbb{Z}_{n_1}^* \times \ldots \times \mathbb{Z}_{n_r}^*\right| = \left|\mathbb{Z}_n^*\right|$$

and therefore,

$$\begin{aligned}\varphi(n_1) \cdots \varphi(n_r) &= \left|\mathbb{Z}_{n_1}^*\right| \cdots \left|\mathbb{Z}_{n_r}^*\right| \\ &= \left|\mathbb{Z}_{n_1}^* \times \ldots \times \mathbb{Z}_{n_r}^*\right| = \left|\mathbb{Z}_n^*\right| = \varphi(n).\end{aligned}$$

Now we are ready to count the cardinality of $\mathbb{Z}_n^*$: let $n = p_1^{a_1} \cdots p_r^{a_r}$ be the prime factorization of n, $p_i \neq p_j$ whenever $i \neq j$. Then

$$\begin{aligned}\varphi(n) &= \varphi(p_1^{a_1}) \cdots \varphi(p_r^{a_r}) = (p_1^{a_1} - p_1^{a_1 - 1}) \cdots (p_r^{a_r} - p_r^{a_r - 1}) \\ &= p_1^{a_1}(1 - \frac{1}{p_1}) \cdots p_r^{a_r}(1 - \frac{1}{p_r}) = n(1 - \frac{1}{p_1}) \cdots (1 - \frac{1}{p_r}).\end{aligned}$$

By expressing the Corollary 9.1.1 on $\mathbb{Z}_n^*$ using the notations of Definition (9.1.3), we have the following corollary.

Corollary 9.1.2. *If* $\gcd(a, n) = 1$, *then* $a^{\varphi(n)} \equiv 1 \pmod{n}$.

This result is known as *Euler's theorem.* If $n = p$ is a prime, then $\varphi(p) = p - 1$ and Euler's theorem can be formulated as

$$a^{p-1} \equiv 1 \pmod{p}. \tag{9.5}$$

Congruence (9.5) is known as *Fermat's little theorem.*

9.1.5 Group Morphisms

Let G and H be two groups written multiplicatively.

Definition 9.1.5. *A mapping* $f : G \to H$ *such that* $f(g_1 g_2) = f(g_1)f(g_2)$ *for all* $g_1, g_2 \in G$ *is called a group morphism from* G *to* H.

It follows that $f(1) = f(1 \cdot 1) = f(1)f(1)$ and, hence, $f(1) = 1$. Moreover, $1 = f(1) = f(gg^{-1}) = f(g)f(g^{-1})$, which implies that $f(g^{-1}) = f(g)^{-1}$. By induction it also follows that $f(g^k) = f(g)^k$ for any integer k.

Definition 9.1.6. *Let* $f : G \to H$ *be a group morphism.*

1. *The set* $\mathrm{Ker}(f) = f^{-1}(1) = \{g \in G \mid f(g) = 1\}$ *is called the kernel of* f.
2. *The set* $\mathrm{Im}(f) = f(G) = \{f(g) \mid g \in G\}$ *is called the image or the range of* f.

If $f(k_1) = f(k_2) = 1$, then also $f(k_1 k_2) = 1$ and $f(k_1^{-1}) = 1$, i.e., $\mathrm{Ker}(f)$ is closed under multiplication and inversion, which means that $\mathrm{Ker}(f)$ is a subgroup of G. In fact, if $k \in \mathrm{Ker}(f)$, then also

$$f(gkg^{-1}) = f(g)f(k)f(g)^{-1} = f(g)f(g)^{-1} = 1,$$

so $\mathrm{Ker}(f)$ is even a normal subgroup. Similarly, it can be seen that $\mathrm{Im}(f)$ is a subgroup of H, not necessarily normal.

A morphism $f : G \to H$ is *injective* if $g_1 \neq g_2$ implies $f(g_1) \neq f(g_2)$, *surjective* if $f(G) = H$ and *bijective* if it is both injective and surjective. Injective, surjective and bijective morphisms are called *monomorphisms*, *epimorphisms*, and *isomorphisms* respectively.

Two groups G and G' are called *isomorphic*, denoted by $G \simeq G'$, if there is an isomorphism $f : G \to G'$. Notice that two isomorphic groups differ only in notations (any element g is replaced with $f(g)$).

An interesting property is given in the following lemma:

Lemma 9.1.6. *Let* $f : G \to H$ *be a group morphism. Then the factor group* $G/\mathrm{Ker}(f)$ *is isomorphic to* $\mathrm{Im}(H)$.

Proof. First notice that $\mathrm{Im}(H)$ is always a subgroup of H and that $\mathrm{Ker}(f)$ is a normal subgroup of G, so the factor group $G/\mathrm{Ker}(f)$ can be defined. We will denote $K = \mathrm{Ker}(f)$ for short and define a function $F : G/K \to \mathrm{Im}(H)$ by $F(gK) = f(g)$. The first thing to be verified is that F is well-defined, i.e, the value of F does not depend on the choice of the coset representative g. But this is straightforward: If $g_1 K = g_2 K$, then by Lemma 9.1.1, $g_1^{-1} g_2 \in K$ and, by the definition of K, $f(g_1^{-1} g_2) = 1$, which implies that $f(g_1) = f(g_2)$ and $F(g_1 K) = f(g_1) = f(g_2) = F(g_2 K)$. It is clear that F is a group morphism. The injectivity can be seen as follows: if $F(g_1 K) = F(g_2 K)$, then by definition, $f(g_1) = f(g_2)$, hence $f(g_1^{-1} g_2) = 1$, which means that $g_1^{-1} g_2 \in K$. But this is to say, by Lemma 9.1.1, that $g_1 K = g_2 K$. It is clear that F is surjective, hence F is an isomorphism. □

9.1.6 Direct Product

Let G and G' be two groups. We can give the Cartesian product $G \times G'$ a group structure by defining

$$(g_1, g_1')(g_2, g_2') = (g_1 g_2, g_1' g_2').$$

It is easy to see that $G \times G'$ becomes a group with $(1, 1')$ as the neutral element[3] and $(g, g') \mapsto (g^{-1}, {g'}^{-1})$ as the inversion. The group $G \times G'$ is called the (outer) *direct product* of G and G'. The direct product is essentially commutative, since $G \times G'$ and $G' \times G$ are isomorphic; an isomorphism is given by $(g, g') \mapsto (g', g)$. Also, $(G_1 \times G_2) \times G_3 \simeq G_1 \times (G_2 \times G_3)$ (isomorphism given by $((g_1, g_2), g_3) \mapsto (g_1, (g_2, g_3))$, so we may as well write this product as $G_1 \times G_2 \times G_3$. This generalizes naturally to the direct products of any number of groups.

If a group G has subgroups H_1 and H_2 such that G is isomorphic to $H_1 \times H_2$, we say that G is the (inner) *direct product* of subgroups H_1 and H_2. We then write also $G = H_1 \times H_2$. Notice also that in this case, the subgroup H_1 (and also H_2) is normal. To see this, we identify, with some abuse of notations, G and $H_1 \times H_2$ and $H_1 = \{(h, 1) \mid h \in H_1\}$. Then

$$(h_1, h_2)(h, 1)(h_1, h_2)^{-1} = (h_1 h h_1^{-1}, 1) \in H_1,$$

as required.

The following lemma will make the notion of the factor group more natural:

Lemma 9.1.7. *If $G = H_1 \times H_2$ (inner) then $G/H_1 \simeq H_2$.*

Proof. Since $G = H_1 \times H_2$, there is an isomorphism $f : G \to H_1 \times H_2$ which gives each $g \in G$ a representation $f(g) = (h_1, h_2)$, where $h_1 \in H_1$ and $h_2 \in H_2$. A mapping $p : H_1 \times H_2 \to H_2$ defined by $p(h_1, h_2) = h_2$ is evidently a surjective morphism, called a *projection* onto H_2. Because f is an isomorphism, the concatenated mapping $pf : G \to H_2$ is also a surjective morphism. Now $p(f(g)) = 1$ if and only if $f(g) = (h_1, 1)$, which happens if and only if $g \in H_1$. This means that $\mathrm{Ker}(pf) = H_1$, so by Lemma 9.1.6 we have $G/H_1 \simeq H_2$. □

Example 9.1.5. Consider again the group $\mathbb{Z}_n^*$, where $n = p_1^{a_1} \cdots p_r^{a_r}$ is a prime decomposition of n. It can be shown that the mapping

$$\mathbb{Z}_{p_1^{a_1}}^* \times \ldots \times \mathbb{Z}_{p_r^{a_r}}^* \to \mathbb{Z}_n^*$$

given by the Chinese Remainder Theorem is an isomorphism.

[3] Here 1 and $1'$ are the neutral elements of G and G' respectively.

9.2 Fourier Transforms

9.2.1 Characters of Abelian Groups

Let G be a finite abelian group written additively. A *character* of G is a morphism $\chi : G \to \mathbb{C} \setminus \{0\}$, i.e., each character satisfies the condition $\chi(g_1 + g_2) = \chi(g_1)\chi(g_2)$ for any elements g_1 and $g_2 \in G$. It follows that $\chi(0) = 1$. Denoting $n = |G|$, we also have by Corollary (9.1.1) that $\chi(g)^n = \chi(ng) = \chi(0) = 1$, so any character value is a nth root of unity.[4]

An interesting property of characters is that for two characters χ_1, χ_2, we can define the *product character* $\chi_1\chi_2 : G \to \mathbb{C}$ by $\chi_1\chi_2(g) = \chi_1(g)\chi_2(g)$. Moreover, it is easy to see that, with respect to this product, the characters also form an abelian group called the *character group* or the *dual group* of G and denoted by $\widehat{G}$. The neutral element χ_0 of the dual group is called the *principal character* or the *trivial character* and is defined by $\chi_0(g) = 1$ for each element $g \in G$. The inverse of a character χ is character χ^{-1} defined by $\chi^{-1}(g) = \chi(g)^{-1}$.

Example 9.2.1. Let us determine the characters of a cyclic group

$$G = \{g, 2g, \ldots, (n-1)g, ng = 0\}.$$

It is easy to see that any cyclic group of order n is isomorphic to $\mathbb{Z}_n$, the additive group of integers modulo n, so we can consider $\mathbb{Z}_n$ as a "prototype" when studying cyclic groups. For any fixed $y \in \mathbb{Z}$ we define mapping $\chi_y : \mathbb{Z} \to \mathbb{C}$ by

$$\chi_y(x) = e^{\frac{2\pi i x y}{n}}.$$

Since $e^{2\pi i} = 1$, χ_y has period n, so we can, in fact, consider χ_y as a mapping $\mathbb{Z}_n \to \mathbb{C}$. Moreover, since $\chi_y(x) = \chi_x(y)$, we can also assume that $y \in \mathbb{Z}_n$ instead of $y \in \mathbb{Z}$. Now

$$\chi_y(x + z) = e^{\frac{2\pi i y(x+z)}{n}} = e^{\frac{2\pi i x y}{n}} e^{\frac{2\pi i z y}{n}} = \chi_y(x)\chi_y(z),$$

which means that each χ_y is, in fact, a character of $\mathbb{Z}_n$. Moreover, if y and z are some representatives of distinct cosets modulo n, then also χ_y and χ_z are different. Namely, if $\chi_y = \chi_z$, then especially $\chi_y(1) = \chi_z(1)$, i.e., $e^{\frac{2\pi i y}{n}} = e^{\frac{2\pi i z}{n}}$, which implies that $y = z + k \cdot n$ for some integer k. But then y and z represent the same coset, a contradiction.

It is straightforward to see that $\chi_a\chi_b = \chi_{a+b}$, so the characters of $\mathbb{Z}_n$ also form a cyclic group of order n, generated by χ_1, for instance. This can be summarized as follows: the character group of $\mathbb{Z}_n$ is isomorphic to $\mathbb{Z}_n$; hence, the character group of any cyclic group is isomorphic to the group itself.

[4] A nth root of unity is a complex number x satisfying $x^n = 1$.

The fact that the dual group of a cyclic group is isomorphic to the group itself, can be generalized: a well-known theorem states that any finite abelian group G can be expressed as a direct sum[5] (or as a direct product, if the group operation is thought as multiplication) of cyclic groups G_i:

$$G = G_1 \oplus \ldots \oplus G_m. \tag{9.6}$$

Lemma 9.2.1. *Let G be an abelian group as above. Then $G \simeq \widehat{G}$.*

Proof. By Example 9.2.1 we know that for each i, $\widehat{G_i} \simeq \widehat{G}_i$, so it suffices to demonstrate that $\widehat{G} = \widehat{G}_1 \times \ldots \times \widehat{G}_m$. For that purpose, let $\chi_1, \ldots, \chi_m$ be some characters of $G_1, \ldots, G_m$. Since $G = G_1 \oplus \ldots \oplus G_m$, each element $g \in G$ can be uniquely expressed as

$$g = g_1 + \ldots + g_m,$$

where $g_i \in G_i$. This allows us to define a function $\chi : G \to \mathbb{C} \setminus \{0\}$ by

$$\chi(g) = \chi_1(g_1) \ldots \chi_m(g_m). \tag{9.7}$$

It is now easy to see that χ is a character of G. Moreover, if $\chi_i' \neq \chi_i$, then there is an element $g_i \in G_i$ such that $\chi_i'(g_i) \neq \chi_i(g_i)$ and, hence,

$$\begin{aligned}\chi'(g_i) &= \chi_1(0) \ldots \chi_i'(g_i) \ldots \chi_m(0) \\ &\neq \chi_1(0) \ldots \chi_i(g_i) \ldots \chi_m(0) = \chi(g_i),\end{aligned}$$

which means that characters of G defined by Equation (9.7) are all different for different choices of $\chi_1, \ldots, \chi_m$.

On the other hand, all the characters of G can be expressed as in (9.7). For, if χ is a character of G, we can define χ_i by restricting χ to G_i. It is easy to see that each χ_i is a character of G_i and that $\chi = \chi_1 \ldots \chi_m$. □

As an application, we will find some important characters.

Example 9.2.2. Consider $\mathbb{F}_2^m$, the m-dimensional vector space over the binary field. Each element in the additive group of $\mathbb{F}_2^m$ has order 2, so group $\mathbb{F}_2^m$ has a simple decomposition:

$$\mathbb{F}_2^m = \underbrace{\mathbb{F}_2 \oplus \ldots \oplus \mathbb{F}_2}_{m \text{ components}}.$$

Now it suffices to determine the characters of $\mathbb{F}_2$, since by Lemma 9.2.1, characters of $\mathbb{F}_2^m$ are just the m-fold products of the characters of $\mathbb{F}_2$. But the characters of $\mathbb{F}_2 = \mathbb{Z}_2$ were already found in Example 9.2.1: They are

$$\chi_y(x) = e^{\frac{2\pi i x y}{2}} = (-1)^{xy}$$

[5] Group G is a direct sum of subgroups $G_1, \ldots, G_m$ if each element of G has unique representation $g = g_1 + \ldots + g_m$, where $g_i \in G_i$.

for $y \in \{0,1\}$. Therefore, any character of $\mathbb{F}_2^m$ can be written as

$$\chi_{\boldsymbol{y}}(\boldsymbol{x}) = (-1)^{x_1 y_1} \cdot \ldots \cdot (-1)^{x_m y_m} = (-1)^{\boldsymbol{x}\cdot\boldsymbol{y}},$$

where $\boldsymbol{x} \cdot \boldsymbol{y} = x_1 y_1 + \ldots + x_m y_m$ is the standard inner product of vectors $\boldsymbol{x} = (x_1, \ldots, x_m)$ and $\boldsymbol{y} = (y_1, \ldots, y_m)$.

9.2.2 Orthogonality of the Characters

Let $G = \{g_1, \ldots, g_n\}$ be a finite abelian group. The functions $f : G \to \mathbb{C}$ clearly form a vector space V over $\mathbb{C}$, addition and scalar multiplication defined pointwise: $(f+h)(g) = f(g) + h(g)$ and $(c \cdot f)(g) = c \cdot f(g)$ (see also Section 9.3). We also have an alternative way of thinking about V: each function $f : G \to \mathbb{C}$ can be considered as an n-tuple

$$(f_{g_1}, \ldots, f_{g_n}) = (f(g_1), \ldots, f(g_n)), \tag{9.8}$$

so the vector space V evidently has dimension n. The natural basis of V is given by $\boldsymbol{e}_1 = (1, 0, \ldots, 0)$, $\boldsymbol{e}_2 = (0, 1, \ldots, 0)$, ..., $\boldsymbol{e}_n = (0, 0, \ldots, 1)$. In other words, $\boldsymbol{e}_i$ is a function $G \to \mathbb{C}$ defined as

$$\boldsymbol{e}_i(g_j) = \begin{cases} 1, \text{ if } i = j \\ 0, \text{ otherwise.} \end{cases}$$

The standard inner product in space V is defined by

$$\langle f \mid h \rangle = \sum_{i=1}^{n} f^*(g_i) h(g_i), \tag{9.9}$$

and the inner product also induces norm in the very natural way:

$$||h|| = \sqrt{\langle h \mid h \rangle}.$$

Basis $\{\boldsymbol{e}_1, \ldots \boldsymbol{e}_n\}$ is clearly orthogonal with respect to the standard inner product. Another orthogonal basis is the characters basis:

Lemma 9.2.2. *If χ_i and χ_j are characters of G, then*

$$\langle \chi_i \mid \chi_j \rangle = \begin{cases} 0, \text{ if } i \neq j \\ n, \text{ if } i = j. \end{cases} \tag{9.10}$$

Proof. First, it is worth noticing that

$$1 = |\chi(g)|^2 = \chi^*(g)\chi(g),$$

which implies that $\chi^*(g) = \chi(g)^{-1}$ for any $g \in G$. Then

$$\langle \chi_i \mid \chi_j \rangle = \sum_{k=1}^{n} \chi_i^*(g_k)\chi_j(g_k) = \sum_{k=1}^{n} \chi_i^{-1}(g_k)\chi_j(g_k) = \sum_{k=1}^{n} (\chi_i^{-1}\chi_j)(g_k).$$

If $i = j$, then $\chi_i^{-1}\chi_j$ is the principal character and the claim for $i = j$ follows immediately. On the other hand, if $i \neq j$, then $\chi = \chi_i^{-1}\chi_j$ is a nontrivial character of G and it suffices to show that

$$S = \sum_{k=1}^{n} \chi(g_k) = 0$$

for a nontrivial character χ. Because of nontriviality, there exists an element $g \in G$ such that $\chi(g) \neq 1$. Furthermore, mapping $g \mapsto g + g_i$ is a permutation of G for any fixed g, so

$$S = \sum_{k=1}^{n} \chi(g_k) = \sum_{k=1}^{n} \chi(g + g_k) = \chi(g) \sum_{k=1}^{n} \chi(g_k) = \chi(g)S.$$

The claim $S = 0$ follows now from the equation $(1 - \chi(g))S = 0$. □

Since the characters are orthogonal and there are $n = |G|$ of them, they also form a basis. By scaling the characters to the unit norm, we obtain an orthonormal basis $\mathcal{B} = \{B_1, \ldots, B_n\}$, where

$$B_i = \frac{1}{\sqrt{n}} \chi_i.$$

Other interesting features of the characters can be easily deduced: let us define a matrix $X \in \mathbb{C}^{n \times n}$ by

$$X_{ij} = \chi_j(g_i).$$

By denoting the transposed complex conjugate of X by X^*, since $X^*_{ij} = \chi_i^*(g_i)$, we get

$$\begin{aligned}
(X^*X)_{ij} &= \sum_{k=1}^{n} X^*_{ik} X_{kj} \\
&= \sum_{k=1}^{n} \chi_i^*(g_k)\chi_j(g_k) = \langle \chi_i \mid \chi_j \rangle \\
&= \begin{cases} 0, \text{ if } i \neq j \\ n, \text{ if } i = j, \end{cases}
\end{aligned}$$

which actually states that $X^*X = nI$, which implies that $X^{-1} = \frac{1}{n}X^*$. But since any matrix commutes with its inverse, we also have $XX^* = nI$, which can be written as

$$(XX^*)_{ij} = \begin{cases} 0, \text{ if } i \neq j \\ n, \text{ if } i = j, \end{cases}$$

or as

$$\sum_{k=1}^{n} \chi_k(g_i)\chi_k^*(g_j) = \begin{cases} 0, \text{ if } i \neq j \\ n, \text{ if } i = j. \end{cases} \tag{9.11}$$

Equations (9.10) and (9.11) are known as the *orthogonality relations* of characters. Notice also that, by choosing χ_j as the principal character in (9.10) and g_j as the neutral element in (9.11), we get a useful corollary:

Corollary 9.2.1.

$$\sum_{k=1}^{n} \chi(g_k) = \begin{cases} n, \textit{ if } \chi \textit{ is the principal character} \\ 0, \textit{ otherwise.} \end{cases} \tag{9.12}$$

$$\sum_{k=1}^{n} \chi_k(g) = \begin{cases} n, \textit{ if } g \textit{ is the neutral element} \\ 0, \textit{ otherwise.} \end{cases} \tag{9.13}$$

In groups $\mathbb{Z}_n$ and $\mathbb{F}_2^m$ we have a nice symmetry: $\chi_i(g_j) = \chi_j(g_i)$. Therefore, the above formulae can be contracted to one formula in both groups. In $\mathbb{Z}_n$ it becomes

$$\sum_{y \in \mathbb{Z}_n} e^{\frac{2\pi i x y}{n}} = \begin{cases} n, \text{ if } x = 0 \\ 0, \text{ otherwise,} \end{cases}$$

and in $\mathbb{F}_2^m$ we have

$$\sum_{\boldsymbol{y} \in \mathbb{F}_2^m} (-1)^{\boldsymbol{x} \cdot \boldsymbol{y}} = \begin{cases} 2^m, \text{ if } \boldsymbol{x} = \boldsymbol{0} \\ 0, \quad \text{otherwise.} \end{cases}$$

9.2.3 Discrete Fourier Transform

We have now all the tools for defining the discrete Fourier transform: any element $f \in V$ (recall that V is the vector space of functions $G \to \mathbb{C}$) has unique representation with respect to basis $\mathcal{B} = \{\frac{1}{\sqrt{n}}\chi_1, \ldots, \frac{1}{\sqrt{n}}\chi_n\}$:

$$f = \widehat{f}_1 B_1 + \ldots + \widehat{f}_n B_n. \tag{9.14}$$

Definition 9.2.1. *The function $\widehat{f} : G \to \mathbb{C}$ defined by*

$$\widehat{f}(g_i) = \widehat{f}_i,$$

where $\widehat{f}_i$ is the coefficient of B_i in the representation (9.14) is called the discrete Fourier transform of f.

Because $\mathcal{B}$ is an orthonormal basis, we can easily extract any coefficient $\widehat{f}_i$ by calculating the inner product of B_i and (9.14):

$$\langle B_i \mid f \rangle = \sum_{j=1}^{n} \langle B_i \mid \widehat{f}_j B_j \rangle = \sum_{j=1}^{n} \widehat{f}_j \langle B_i \mid B_j \rangle = \widehat{f}_i.$$

Thus, the Fourier transform can be written as

$$\widehat{f}(g_i) = \langle B_i \mid f \rangle = \frac{1}{\sqrt{n}} \sum_{k=1}^{n} \chi_i^*(g_k) f(g_k). \tag{9.15}$$

By its very definition, it is clear that $\widehat{f+h} = \widehat{f} + \widehat{h}$ and $\widehat{cf} = c\widehat{f}$ for any functions f, h and any $c \in \mathbb{C}$.

Example 9.2.3. An interesting corollary of the orthogonality relation (9.11) can be easily derived:

$$\begin{aligned}
||\widehat{f}||^2 = \langle \widehat{f} \mid \widehat{f} \rangle &= \sum_{i=1}^{n} \widehat{f}^*(g_i)\widehat{f}(g_i) \\
&= \sum_{i=1}^{n} \frac{1}{\sqrt{n}} \sum_{k=1}^{n} \chi_i(g_k) f^*(g_k) \frac{1}{\sqrt{n}} \sum_{l=1}^{n} \chi_i^*(g_l) f(g_l) \\
&= \frac{1}{n} \sum_{k=1}^{n} \sum_{l=1}^{n} f^*(g_k) f(g_l) \sum_{i=1}^{n} \chi_i(g_k) \chi_i^*(g_l) \\
&= \frac{1}{n} \sum_{k=1}^{n} f^*(g_k) f(g_k) n = \langle f \mid f \rangle = ||f||^2.
\end{aligned}$$

The equation $||\widehat{f}|| = ||f||$ thus obtained is known as *Parseval's identity.*

Example 9.2.4. Let $G = \mathbb{Z}_n$. As we saw in Example (9.2.1), the characters of $\mathbb{Z}_n$ are all of form

$$\chi_y(x) = e^{\frac{2\pi i x y}{n}},$$

and, therefore, the Fourier transform of a function $f : \mathbb{Z}_n \to \mathbb{C}$ takes form

$$\widehat{f}(x) = \frac{1}{\sqrt{n}} \sum_{y \in \mathbb{Z}_n} e^{-\frac{2\pi i x y}{n}} f(y).$$

Example 9.2.5. The characters of group $G = \mathbb{F}_2^m$ are

$$\chi_{\boldsymbol{y}}(\boldsymbol{x}) = (-1)^{\boldsymbol{x} \cdot \boldsymbol{y}},$$

so the Fourier transform of $f : \mathbb{F}_2^m \to \mathbb{C}$ looks like

$$\widehat{f}(\boldsymbol{x}) = \frac{1}{\sqrt{2^m}} \sum_{\boldsymbol{y} \in \mathbb{F}_2^m} (-1)^{\boldsymbol{x} \cdot \boldsymbol{y}} f(\boldsymbol{y}).$$

This Fourier transform in $\mathbb{F}_2^m$ is also called a *Hadamard transform, Walsh transform,* or *Hadamard-Walsh transform.*

9.2.4 The Inverse Fourier Transform

Notice that Equation (9.15) can be rewritten as

$$\begin{pmatrix} \widehat{f}(g_1) \\ \widehat{f}(g_2) \\ \vdots \\ \widehat{f}(g_n) \end{pmatrix} = \frac{1}{\sqrt{n}} \begin{pmatrix} \chi_1^*(g_1) & \chi_1^*(g_2) & \dots & \chi_1^*(g_n) \\ \chi_2^*(g_1) & \chi_2^*(g_2) & \dots & \chi_2^*(g_n) \\ \vdots & \vdots & \ddots & \vdots \\ \chi_n^*(g_1) & \chi_n^*(g_2) & \dots & \chi_n^*(g_n) \end{pmatrix} \begin{pmatrix} f(g_1) \\ f(g_2) \\ \vdots \\ f(g_n) \end{pmatrix}, \tag{9.16}$$

and that the matrix appearing in the right-hand side of (9.16) is X^*, the transpose complex conjugate of matrix X defined as $X_{ij} = \chi_j(g_i)$.[6] By multiplying (9.16) by X we get

$$\begin{pmatrix} f(g_1) \\ f(g_2) \\ \vdots \\ f(g_n) \end{pmatrix} = \frac{1}{\sqrt{n}} \begin{pmatrix} \chi_1(g_1) & \chi_2(g_1) & \dots & \chi_n(g_1) \\ \chi_1(g_2) & \chi_2(g_2) & \dots & \chi_n(g_2) \\ \vdots & \vdots & \ddots & \vdots \\ \chi_1(g_n) & \chi_2(g_n) & \dots & \chi_n(g_n) \end{pmatrix} \begin{pmatrix} \widehat{f}(g_1) \\ \widehat{f}(g_2) \\ \vdots \\ \widehat{f}(g_n) \end{pmatrix}, \tag{9.17}$$

which gives the idea for the following definition.

Definition 9.2.2. *Let $f : G \to \mathbb{C}$ be a function. The inverse Fourier transform of f is defined to be*

$$\widetilde{f}(g_i) = \frac{1}{\sqrt{n}} \sum_{k=1}^{n} \chi_k(g_i) f(g_k). \tag{9.18}$$

Keeping equations (9.16) and (9.17) in mind, it is clear that $\widetilde{\widehat{f}} = \widehat{\widetilde{f}} = f$.

Example 9.2.6. In $\mathbb{Z}_n$ we have $\chi_x(y) = \chi_y(x)$, so the inverse Fourier transform is quite symmetric to the Fourier transform:

$$\widetilde{f}(x) = \frac{1}{\sqrt{n}} \sum_{y \in \mathbb{Z}_n} e^{\frac{2\pi i x y}{n}} f(y).$$

In group $\mathbb{F}_2^m$ the symmetry is perfect:

$$\widetilde{f}(\boldsymbol{x}) = \frac{1}{\sqrt{2^m}} \sum_{\boldsymbol{y} \in \mathbb{F}_2^m} (-1)^{\boldsymbol{x} \cdot \boldsymbol{y}} f(\boldsymbol{y}) = \widehat{f}(\boldsymbol{x}).$$

---

[6] Equation (9.16) makes Parseval's identity even clearer: matrices $\frac{1}{\sqrt{n}}X$ and $\frac{1}{\sqrt{n}}X^*$ are unitary and, therefore, they preserve the norm in V.

9.2.5 Fourier Transform and Periodicity

We will now give an illustration on how powerfully the Fourier transform can extract information about the periodicity.

Let $f : G \to \mathbb{C}$ be a function with *period* $p \in G$, i.e., $f(g+p) = f(g)$ for any $g \in G$. Then

$$\begin{aligned}\widehat{f}(g_i) &= \frac{1}{\sqrt{n}} \sum_{k=1}^{n} \chi_i^*(g_k) f(g_k) \\ &= \frac{1}{\sqrt{n}} \sum_{k=1}^{n} \chi_i(g_k + p - p)^* f(g_k + p) \\ &= \chi_i^*(-p) \frac{1}{\sqrt{n}} \sum_{k=1}^{n} \chi_i^*(g_k + p) f(g_k + p) \\ &= \chi_i^*(-p) \widehat{f}(g_i),\end{aligned}$$

which implies that $\widehat{f}(g_i) = 0$ whenever $\chi_i^*(-p) = \chi_i(-p)^{-1}\chi_i(p) \neq 1$.

Example 9.2.7. In group $\mathbb{Z}_n$ the above equation takes form

$$\widehat{f}(x) = \mathrm{e}^{-\frac{2\pi i x p}{n}} \widehat{f}(x),$$

which means that $\widehat{f}(x) = 0$ whenever $\mathrm{e}^{-\frac{2\pi i x p}{n}} \neq 1$, which happens exactly when xp is not divisible by n. If also $\gcd(p, n) = 1$, then $-xp$ can be divisible by n and is so if and only if x is.

9.3 Linear Algebra

9.3.1 Preliminaries

A *vector space* over complex numbers is an abelian group V (usually written additively) that is equipped with *scalar multiplication*, which is a mapping $\mathbb{C} \times V \to V$. The scalar multiplication is usually denoted by $(c, \boldsymbol{x}) \mapsto c\boldsymbol{x}$. It is required that the following *vector space axioms* must be fulfilled for all $c, c_1, c_2 \in \mathbb{C}$ and $\boldsymbol{x}, \boldsymbol{x}_1, \boldsymbol{x}_2 \in V$:

1. $c_1(c_2\boldsymbol{x}) = (c_1c_2)\boldsymbol{x}$.
2. $(c_1 + c_2)\boldsymbol{x} = c_1\boldsymbol{x} + c_2\boldsymbol{x}$.
3. $c(\boldsymbol{x}_1 + \boldsymbol{x}_2) = c\boldsymbol{x}_1 + c\boldsymbol{x}_2$.
4. $1\boldsymbol{x} = \boldsymbol{x}$.

Again, to be precise, we should talk about vector space $(V, +, \mathbf{0}, -, \mathbb{C}, \cdot)$ instead of space V, but to simplify the notations we identify the space and the underlying set V. The elements of V are called *vectors*.

Example 9.3.1. Let $V = \mathbb{C}^n$, the set of n-tuples over complex numbers. Set $\mathbb{C}^n$ equipped with sum

$$(x_1, \ldots, x_n) + (y_1, \ldots, y_n) = (x_1 + y_1, \ldots, x_n + y_n),$$

zero element $(0, \ldots, 0)$ and inversion

$$(x_1, \ldots, x_n) \mapsto (-x_1, \ldots, -x_n)$$

is evidently an abelian group. This set equipped with scalar multiplication

$$c(x_1, \ldots, x_n) = (cx_1, \ldots, cx_n)$$

becomes a vector space, as is easily verified.

Example 9.3.2. The set of functions $\boldsymbol{x} : \mathbb{N} \to \mathbb{C}$, for which the series

$$\sum_{n=1}^{\infty} |\boldsymbol{x}(n)|^2$$

is convergent, also constitutes a vector space over $\mathbb{C}$. The sum and scalar product are again defined pointwise: for functions $\boldsymbol{x}$ and $\boldsymbol{y}$, $(\boldsymbol{x} + \boldsymbol{y})(n) = \boldsymbol{x}(n) + \boldsymbol{y}(n)$ and $(c\boldsymbol{x})(n) = c\boldsymbol{x}(n)$. To simplify the notations, we write also $x_n = \boldsymbol{x}(n)$. To verify that the sum of two elements stays in the space, we also have to check that the series $\sum_{n=1}^{\infty} |x_n + y_n|^2$ is convergent whenever $\sum_{n=1}^{\infty} |x_n|^2$ and $\sum_{n=1}^{\infty} |y_n|^2$ are convergent. For this purpose, we can use estimations

$$\begin{aligned} |x + y|^2 &= |x|^2 + 2\mathrm{Re}(x^* y) + |y|^2 \\ &\le |x|^2 + 2|x||y| + |y|^2 \le 2|x|^2 + 2|y|^2 \end{aligned}$$

for each summand. The vector space of this example is denoted by $L_2(\mathbb{C})$. Notice that the domain $\mathbb{N}$ in the definition could be replaced with any numerable set. If, instead of $\mathbb{N}$, there is a finite domain $\{1, \ldots, n\}$, the convergence condition would be unnecessary and the space would become essentially the same as $\mathbb{C}^n$.

Definition 9.3.1. *A linear combination of vectors* $\boldsymbol{x}_1, \ldots, \boldsymbol{x}_n$ *is a finite sum* $c_1\boldsymbol{x}_1 + \ldots + c_n\boldsymbol{x}_n$. *A set* $S \subset V$ *is called linearly independent if*

$$c_1\boldsymbol{x}_1 + \ldots + c_n\boldsymbol{x}_n = \mathbf{0}$$

implies $c_1 = \ldots = c_n = 0$ *whenever* $\boldsymbol{x}_1, \ldots, \boldsymbol{x}_n \in S$. *A set that is not linearly independent is linearly dependent.*

Definition 9.3.2. *For any set* $S \subseteq V$, $L(S)$ *is the set of all linear combinations of vectors in* S. *By its very definition, set* $L(S)$ *is a vector space contained in* V. *We say that* $L(S)$ *is generated by* S *and also call* $L(S)$ *the span of* S.

The proof of the following lemma is left as an exercise.

Lemma 9.3.1. *A set $S \subseteq V$ is linearly dependent if and only if some vector $\boldsymbol{x} \in S$ can be represented as a linear combination of vectors in $S \setminus \{\boldsymbol{x}\}$.*

In ligth of the previous lemma, a linearly dependent set contains some "unnecessary" elements in the sense that they are not needed when making linear combinations. For, if $\boldsymbol{x} \in S$ can be expressed as a linear combination of $S \setminus \{\boldsymbol{x}\}$, then clearly $L(S \setminus \{\boldsymbol{x}\}) = L(S)$ (cf. Exercise 1).

Definition 9.3.3. *A set $B \subset V$ is a basis of V if $V = L(B)$ and B is linearly independent.*

It can be shown that all the bases have the same cardinality (See Exercise 2). This gives the following definition.

Definition 9.3.4. *The dimension* $\dim(V)$ *of a vector space V is the cardinality of a basis of V.*

Example 9.3.3. Vectors $\boldsymbol{e}_1 = (1, 0, \ldots, 0)$, $\boldsymbol{e}_2 = (0, 1, \ldots, 0)$, ..., $\boldsymbol{e}_n = (0, 0, \ldots, 1)$ form a basis of $\mathbb{C}^n$. This is a so-called *natural basis.* Thus $\mathbb{C}^n$, has dimension n.

Example 9.3.4. As in the previous example, we could define $\boldsymbol{e}_i \in L_2(\mathbb{C})$ by

$$\boldsymbol{e}_i(n) = \begin{cases} 1, \text{ if } i = n \\ 0, \text{ if } i \neq n \end{cases}$$

for each $i \in \mathbb{N}$. It is clear that the set $\mathcal{E} = \{\boldsymbol{e}_1, \boldsymbol{e}_2, \boldsymbol{e}_3, \ldots\}$ is linearly independent, but it is *not* a basis of $L_2(\mathbb{C})$ in the sense of definition (9.3.3), because there are vectors in $L_2(\mathbb{C})$ that cannot be expressed as a linear combination of vectors in $\mathcal{E}$. This is because, according to definition (9.3.1), a linear combination is a *finite* sum of vectors (Exercise 4).

Definition 9.3.5. *A subset $W \subseteq V$ is a subspace of V if W is a subgroup of V that is closed under scalar multiplication.*

Example 9.3.5. Consider the n-dimensional vector space $\mathbb{C}^n$. Set

$$W = \{(c_1, \ldots, c_{n-1}, 0) \in \mathbb{C}^n\}$$

is clearly a subspace of $\mathbb{C}^n$ having $\boldsymbol{e}_1$, ... $\boldsymbol{e}_{n-1}$ as basis. Hence, $\dim(W) = n - 1$.

Example 9.3.6. Let

$$W = \{\boldsymbol{x} \in L_2(\mathbb{C}) \mid x_n \neq 0 \text{ only for finitely many } n \in \mathbb{N}\}.$$

It is straightforward to see that W is a subspace of $L_2(\mathbb{C})$ and that the set $\mathcal{E}$ of example (9.3.4) is a basis of W.

9.3.2 Inner Product

A natural way to introduce geometry into a complex vector space V (a vector space over $\mathbb{C}$) is to define the *inner product.*

Definition 9.3.6. *An inner product on V is a mapping $V \times V \to \mathbb{C}$, $(\boldsymbol{x}, \boldsymbol{y}) \mapsto \langle \boldsymbol{x} \mid \boldsymbol{y} \rangle$ that satisfies the following conditions for any $c \in \mathbb{C}$ and any $\boldsymbol{x}$, $\boldsymbol{y}$, and $\boldsymbol{z} \in V$*

1. $\langle \boldsymbol{x} \mid \boldsymbol{y} \rangle = \langle \boldsymbol{y} \mid \boldsymbol{x} \rangle^*$.
2. $\langle \boldsymbol{x} \mid \boldsymbol{x} \rangle \geq 0$ *and* $\langle \boldsymbol{x} \mid \boldsymbol{x} \rangle = 0$ *if and only if* $\boldsymbol{x} = \boldsymbol{0}$.
3. $\langle \boldsymbol{x} \mid c_1 \boldsymbol{y} + c_2 \boldsymbol{z} \rangle = c_1 \langle \boldsymbol{x} \mid \boldsymbol{y} \rangle + c_2 \langle \boldsymbol{x} \mid \boldsymbol{z} \rangle$.

In axiom 1 and hereafter, * means complex conjugation. Notice that by axiom 1 it follows that $\langle \boldsymbol{x} \mid \boldsymbol{x} \rangle$ is always real, so axiom 2 makes sense. From axioms 1 and 3 it follows that

$$\langle c_1 \boldsymbol{x}_1 + c_2 \boldsymbol{x}_2 \mid \boldsymbol{x} \rangle = c_1^* \langle \boldsymbol{x}_1 \mid \boldsymbol{x} \rangle + c_2^* \langle \boldsymbol{x}_2 \mid \boldsymbol{x} \rangle.$$

A vector space equipped with an inner product is also called an *inner product space.*

Example 9.3.7. For vectors $\boldsymbol{x} = (x_1, \ldots, x_n)$, and $\boldsymbol{y} = (y_1, \ldots, y_n)$ in $\mathbb{C}^n$ the formula

$$\langle \boldsymbol{x} \mid \boldsymbol{y} \rangle = x_1^* y_1 + \ldots + x_n^* y_n$$

defines an inner product, as is easily verified. An inner product of $\boldsymbol{x}$ and $\boldsymbol{y} \in L_2(\mathbb{C})$ can be defined by formula

$$\sum_{n=1}^{\infty} x_n^* y_n, \tag{9.19}$$

since the series (9.19) is convergent. This is because

$$\left| \sum_{n=N}^{M} x_n^* y_n \right| \leq \sum_{n=N}^{M} |x_n^* y_n| \leq \sum_{n=N}^{M} \frac{1}{2}(|x_n|^2 + |y_n|^2)$$

and the series $\sum_{n=1}^{\infty} |x_n|^2$ and $\sum_{n=1}^{\infty} |x_n|^2$ converge by the very definition of $L_2(\mathbb{C})$.

Definition 9.3.7. *Vectors $\boldsymbol{x}$ and $\boldsymbol{y}$ are orthogonal if $\langle \boldsymbol{x} \mid \boldsymbol{y} \rangle = 0$. Two subsets $S_1 \subseteq V$ and $S_2 \subseteq V$ are mutually orthogonal if all $\boldsymbol{x}_1 \in S_1$ and $\boldsymbol{x}_2 \in S_2$ are orthogonal. A set $S \subseteq V$ is orthogonal if $\boldsymbol{x}_1 \in S$ and $\boldsymbol{x}_2 \in S$ are orthogonal whenever $\boldsymbol{x}_1 \neq \boldsymbol{x}_2$*

An orthogonal set that does not contain the zero vector $\mathbf{0}$ is always linearly independent: for, if S is such a set and

$$c_1\boldsymbol{x}_1 + \ldots + c_n\boldsymbol{x}_n = \mathbf{0}$$

is a linear combination of vectors in S, then $c_i\langle \boldsymbol{x}_i \mid \boldsymbol{x}_i\rangle = \langle \boldsymbol{x}_i \mid c_i\boldsymbol{x}_i\rangle = \langle \boldsymbol{x}_i \mid \mathbf{0}\rangle = 0$, which implies $c_i = 0$ since $\boldsymbol{x}_i \neq \mathbf{0}$.

Example 9.3.8. Let $W \subseteq V$ be a subspace. Then also

$$W^{\perp} = \{\boldsymbol{x} \in V \mid \langle \boldsymbol{x} \mid \boldsymbol{y}\rangle = 0 \text{ for all } \boldsymbol{y} \in W\}$$

is a subspace of V, the so-called *orthogonal complement* of W.

Lemma 9.3.2 (Cauchy-Schwartz inequality). *For any vectors $\boldsymbol{x}, \boldsymbol{y} \in V$,*

$$|\langle \boldsymbol{x} \mid \boldsymbol{y}\rangle|^2 \leq \langle \boldsymbol{x} \mid \boldsymbol{x}\rangle\langle \boldsymbol{y} \mid \boldsymbol{y}\rangle.$$

Proof. If $\boldsymbol{y} = \mathbf{0}$, then $\langle \boldsymbol{x} \mid \mathbf{0}\rangle = \langle \boldsymbol{x} \mid \mathbf{0} + \mathbf{0}\rangle = \langle \boldsymbol{x} \mid \mathbf{0}\rangle + \langle \boldsymbol{x} \mid \mathbf{0}\rangle$, so $\langle \boldsymbol{x} \mid \mathbf{0}\rangle = 0$ and the claim holds with equality. If $\boldsymbol{y} \neq \mathbf{0}$, we can define $\lambda = -\frac{\langle \boldsymbol{y} | \boldsymbol{x}\rangle}{\langle \boldsymbol{y} | \boldsymbol{y}\rangle}$. Then, by the inner product axiom 2,

$$0 \leq \langle \boldsymbol{x} + \lambda\boldsymbol{y} \mid \boldsymbol{x} + \lambda\boldsymbol{y}\rangle = \langle \boldsymbol{x} \mid \boldsymbol{x}\rangle + \lambda\langle \boldsymbol{x} \mid \boldsymbol{y}\rangle + \lambda^*\langle \boldsymbol{y} \mid \boldsymbol{x}\rangle + |\lambda|^2 \langle \boldsymbol{y} \mid \boldsymbol{y}\rangle,$$

which can also be written as

$$0 \leq \langle \boldsymbol{x} \mid \boldsymbol{x}\rangle - \frac{\langle \boldsymbol{y} \mid \boldsymbol{x}\rangle}{\langle \boldsymbol{y} \mid \boldsymbol{y}\rangle}\langle \boldsymbol{x} \mid \boldsymbol{y}\rangle,$$

and the claim follows. □

Definition 9.3.8. *A norm on a vector space is a mapping $V \to \mathbb{R}$, $\boldsymbol{x} \mapsto ||\boldsymbol{x}||$ such that, for all vectors $\boldsymbol{x}$ and $\boldsymbol{y}$, and $c \in \mathbb{C}$ we have*

1. $||\boldsymbol{x}|| \geq 0$ *and* $||\boldsymbol{x}|| = 0$ *if and only if* $\boldsymbol{x} = \mathbf{0}$.
2. $||c\boldsymbol{x}|| = |c|\,||\boldsymbol{x}||$.
3. $||\boldsymbol{x} + \boldsymbol{y}|| \leq ||\boldsymbol{x}|| + ||\boldsymbol{y}||$.

A vector space equipped with a norm is also called a *normed space* . Any inner product always induces a norm by

$$||\boldsymbol{x}|| = \sqrt{\langle \boldsymbol{x} \mid \boldsymbol{x}\rangle},$$

so an inner product space is always a normed space as well. To verify that $\sqrt{\langle \boldsymbol{x} \mid \boldsymbol{x}\rangle}$ is a norm, it is easy to check that the conditions 1 and 2 are fulfilled; and for condition 3, we use the Cauchy-Schwartz inequality, which can also be stated as $|\langle \boldsymbol{x} \mid \boldsymbol{y}\rangle| \leq ||\boldsymbol{x}||||\boldsymbol{y}||$. Thus,

$$\begin{aligned}
||\boldsymbol{x} + \boldsymbol{y}||^2 &= \langle \boldsymbol{x} + \boldsymbol{y} \mid \boldsymbol{x} + \boldsymbol{y}\rangle = ||\boldsymbol{x}||^2 + 2\mathrm{Re}\langle \boldsymbol{x} \mid \boldsymbol{y}\rangle + ||\boldsymbol{y}||^2 \\
&\leq ||\boldsymbol{x}||^2 + 2\,|\langle \boldsymbol{x} \mid \boldsymbol{y}\rangle| + ||\boldsymbol{y}||^2 \leq ||\boldsymbol{x}||^2 + 2||\boldsymbol{x}||||\boldsymbol{y}|| + ||\boldsymbol{y}||^2 \\
&= (||\boldsymbol{x}|| + ||\boldsymbol{y}||)^2.
\end{aligned}$$

Any norm can be used to define the *distance* on V:

$$d(\boldsymbol{x}, \boldsymbol{y}) = ||\boldsymbol{x} - \boldsymbol{y}||.$$

One can easily verify that $d : V \times V \to \mathbb{R}$ satisfies the axioms of a *distance function*:

For each $\boldsymbol{x}$, $\boldsymbol{y}$, and $\boldsymbol{z} \in V$,

1 $d(\boldsymbol{x}, \boldsymbol{y}) = d(\boldsymbol{y}, \boldsymbol{x}) \geq 0$.
2 $d(\boldsymbol{x}, \boldsymbol{y}) = 0$ if and only if $\boldsymbol{x} = \boldsymbol{y}$.
3 $d(\boldsymbol{x}, \boldsymbol{z}) \leq d(\boldsymbol{x}, \boldsymbol{y}) + d(\boldsymbol{y}, \boldsymbol{z})$.

A characteristic property of a normed space that is also an inner product space is that

$$||\boldsymbol{x} + \boldsymbol{y}||^2 + ||\boldsymbol{x} - \boldsymbol{y}||^2 = 2||\boldsymbol{x}||^2 + 2||\boldsymbol{y}||^2 \tag{9.20}$$

holds for any $\boldsymbol{x}, \boldsymbol{y} \in V$ (See Exercise 5). Equation (9.20) is called the *Parallelogram rule.*

Definition 9.3.9. *Let V be a normed space. If, for each sequence of vectors $\boldsymbol{x}_1$, $\boldsymbol{x}_2$, ... such that*

$$\lim_{m,n \to \infty} ||\boldsymbol{x}_m - \boldsymbol{x}_n|| = 0,$$

there exists a vector $\boldsymbol{x}$ such that

$$\lim_{m \to \infty} ||\boldsymbol{x}_n - \boldsymbol{x}|| = 0,$$

we say that V is complete.

A class of vectors space having great importance in quantum physics and in quantum computation is introduced in the following definition.

Definition 9.3.10. *An inner product space V is a Hilbert space if V is complete with respect to the norm induced by the inner product.*

Example 9.3.9. Both $\mathbb{C}^n$ and $L_2(\mathbb{C})$ are Hilbert spaces, but the subspace W of $L_2(\mathbb{C})$ in Example 9.3.4 is not (cf. Exercise 6).

Definition 9.3.11. *A vector space V is a direct sum of subspaces W_1 and W_2, denoted by $V = W_1 \oplus W_2$, if V is a direct sum of the subgroups W_1 and W_2.*

The next lemma shows that Hilbert spaces are structurally well-behaving:

Lemma 9.3.3. *If a subspace W of a Hilbert space V is itself W a Hilbert space, then $V = W \oplus W^{\perp}$ (see Example (9.3.8)).*

Definition 9.3.12. *Let V and W be complex vector spaces. A mapping $f : V \to W$ is a vector space morphism if*

$$f(c_1\boldsymbol{x} + c_2\boldsymbol{y}) = c_1 f(\boldsymbol{x}) + c_2 f(\boldsymbol{y})$$

for each $\boldsymbol{x}$, $\boldsymbol{y} \in V$ and c_1, $c_2 \in \mathbb{C}$. Vector space morphisms are also called linear mappings or operators.

9.4 Number Theory

9.4.1 Euclid's Algorithm

It should be emphasized that the proof of Lemma 9.1.5 already supplies a recursive method for computing the greatest common divisor of two given natural numbers:

$$\gcd(x, y) = \begin{cases} \gcd(x-y, y), & \text{if } x > y \\ \gcd(x, y-x), & \text{if } x < y \\ x, & \text{if } x = y. \end{cases}$$

It is clear that this method always gives $\gcd(x, y)$ correctly. However, the method is not quite efficient: for instance, in computing $\gcd(x, 1)$ it proceeds as

$$\gcd(x, 1) = \gcd(x-1, 1) = \gcd(x-2, 1) = \ldots = \gcd(1, 1) = 1,$$

so it takes x recursive calls to perform this computation. Keeping in mind that the decimal representation of x has length $\ell(x) = \lfloor \log_{10} x \rfloor$, the computational time is proportional to $10^{\ell(x)}$, i.e., the current algorithm is exponential in $\ell(x)$.

A more efficient method is given by *Euclid's algorithm*, whose core is given in the following lemma.

Lemma 9.4.1. *Given natural numbers $x > y$, there are unique integers d and r such that $x = dy + r$ where $0 \leq r < y$. Numbers y and d can be found by using $\ell(x)^2$ elementary arithmetic operations (multiplication, addition, and subtraction of two digits).*

Euclid's algorithm is based on consecutive application of the previous lemma: given $x > y$, we can find representations

$$x = d_1 y + r_1, \tag{9.21}$$
$$y = d_2 r_1 + r_2, \tag{9.22}$$
$$r_1 = d_3 r_2 + r_3,$$
$$\ldots$$
$$r_{n-2} = d_n r_{n-1} + r_n, \tag{9.23}$$
$$r_{n-1} = d_{n+1} r_n. \tag{9.24}$$

The above procedure certainly terminates with some $r_{n+1} = 0$, since r_1, r_2, ... is a strictly descending sequence of non-negative numbers. We claim that $\gcd(x, y)$ equals to r_n, the last nonzero remainder in this procedure. First, by (9.24), r_n divides r_{n-1}. By (9.23) r_n also divides r_{n-2}. Continuing this way, we see that r_n divides all the numbers r_i and also x and y. Therefore, r_n is a common divisor of x and y. Assume, then, that there is another number s that divides both x and y. By (9.21), s divides r_1. By (9.22), s divides

r_2 and again continuing this reasoning, we see that s eventually divides r_n. Therefore r_n is the greatest common divisor of x and y.

How rapid is this method? Since each $r_i < x$, we know that each iteration step can be done in a time proportional to $\ell(x)^2$. Thus, it suffices to analyze how many times we need to iterate in order to reach $r_{n+1} = 0$. Because one single iteration step gives an equation

$$r_{i-2} = d_i r_{i-1} + r_i$$

where $0 \le r_i < r_{i-1}$, we have $r_i = r_{i-2} - d_i r_{i-1} < r_{i-2} - d_i r_i$. The estimation $r_{i-2} > (1 + d_i) r_i \ge 2r_i$ follows easily. Therefore,

$$x > y > r_1 > 2r_3 > 4r_5 \ldots > 2^i r_{2i+1}.$$

If n is odd, we have $x > 2^{\frac{n-1}{2}} r_n \ge 2^{\frac{n-1}{2}}$. For an even n, we also have $x > 2^{\frac{n-2}{2}} r_{n-1} > 2^{\frac{n-2}{2}} r_n$, so the inequality $x > 2^{\frac{n-2}{2}}$ holds in any case. It follows that $\ell(x) = \lfloor \log_{10} x \rfloor > \log_{10} x - 1 > \frac{n-2}{2} \log_{10} 2 - 1$. Therefore, $n < \frac{2}{\log_{10} 2}(\ell(x) + 1) + 2 = O(\ell(n))$. As a conclusion, we have that, for numbers $x > y$, $\gcd(x, y)$ can be computed using $O(\ell(x)^3)$ elementary arithmetic operations.

9.4.2 Continued Fractions

Example 9.4.1. Let us apply Euclid's algorithm on pair $(263, 189)$:

$$\begin{aligned}
263 &= 1 \cdot 189 + 74,\\
189 &= 2 \cdot 74 + 41,\\
74 &= 1 \cdot 41 + 33,\\
41 &= 1 \cdot 33 + 8,\\
33 &= 4 \cdot 8 + 1,
\end{aligned}$$

and the algorithm terminates at the next round giving $\gcd(263, 189) = 1$. By divisions we obtain

$$\begin{aligned}
\frac{263}{189} &= 1 + \frac{74}{189},\\
\frac{189}{74} &= 2 + \frac{41}{74},\\
\frac{74}{41} &= 1 + \frac{33}{41},\\
\frac{41}{33} &= 1 + \frac{8}{33},\\
\frac{33}{8} &= 4 + \frac{1}{8},
\end{aligned}$$

a set of equations such that the fraction occurring on the right-hand side is the inverse of the fraction on the left-hand side of the next equation. We can thus combine these equations to get a representation

$$\frac{263}{189} = 1 + \cfrac{1}{2 + \cfrac{1}{1 + \cfrac{1}{1 + \cfrac{1}{4 + \cfrac{1}{8}}}}} \tag{9.25}$$

Expression (9.25) is an example of a *finite continued fraction.*

It is clear that this procedure can be done for any positive rational number $\alpha = \frac{p}{q} \geq 1$ in the lowest terms.

For other values of α, there is a unique way to express $\alpha = a_0 + \beta$ where $a_0 \in \mathbb{Z}$ and $\beta \in [0, 1)$. If $\beta \neq 0$, this can also be written as

$$\alpha = a_0 + \frac{1}{\alpha_1}, \qquad \text{where } \alpha_1 = \frac{1}{\beta} > 1.$$

By applying the same procedure recursively to α_1, we get an expansion

$$\alpha = a_0 + \cfrac{1}{a_1 + \cfrac{1}{a_2 + \cfrac{1}{a_3 + \cfrac{1}{\ddots}}}} \tag{9.26}$$

where $a_0 \in \mathbb{Z}$, and a_1, a_2, ... are natural numbers. If α is an irrational number, then the sequence a_0, a_1, a_2, ... never stops (otherwise α would be a rational number). Mainly for typographical reasons, we write (9.26) as

$$\alpha = [a_0, a_1, a_2, \ldots] \qquad a_0 \in \mathbb{Z},\ a_i \in \mathbb{N},\ \text{for } i \geq 1, \tag{9.27}$$

and say that (9.27) is the *continued fraction expansion* of α. It is clear by its very construction that each irrational α has unique infinite continued fraction expansion (9.27).

On the other hand, all rational numbers r have a finite continued fraction expansion

$$r = [a_0, a_1, \ldots, a_n], \qquad a_0 \in \mathbb{Z},\ a_i \in \mathbb{N} \text{ for } 1 \leq n \tag{9.28}$$

that can be found by using Euclid's algorithm. But the expansion (9.28) is never unique, as is shown by the following lemma.

Lemma 9.4.2. *If r has expansion (9.28) of odd length, the r also has expansion (9.28) of even length and vice versa.*

Proof. This follows straightforwardly from the identity

$$[a_0, a_1, \ldots, a_{n-1}, a_n] = [a_0, a_1, \ldots, a_{n-1} + \frac{1}{a_n}].$$

If $a_n \geq 2$, then $[a_0, a_1, \ldots, a_n] = [a_0, a_1 \ldots, a_n - 1, 1]$. If $a_n = 1$, then $[a_0, a_1, \ldots, a_{n-1}, 1] = [a_0, a_1, \ldots, a_{n-2}, a_{n-1} + 1]$. □

Example 9.4.2. The expansion (9.25) can be written in two ways:

$$\frac{263}{189} = 1 + \cfrac{1}{2 + \cfrac{1}{1 + \cfrac{1}{1 + \cfrac{1}{4 + \cfrac{1}{8}}}}} = 1 + \cfrac{1}{2 + \cfrac{1}{1 + \cfrac{1}{1 + \cfrac{1}{4 + \cfrac{1}{7 + \cfrac{1}{1}}}}}}$$

which illustrates how we can either lengthen or shorten any finite continued fraction expansion by 1. It can, however, be shown that if a finite continued fraction expansion is required to end up by $a_n > 1$, then the representation is unique.

Clearly, each finite continued fraction represents some rational number. But even though we now know that each irrational number has unique representation as an infinite continued fraction, we cannot tell a priori if each sequence

$$a_0, a_1, a_2, \ldots, \quad \text{where } a_0 \in \mathbb{Z} \text{ and } a_i \in \mathbb{N}, \text{ when } i \geq 1 \tag{9.29}$$

regarded as a continued fraction represents an irrational number. In other words, we do not yet know if the limit

$$\lim_{n\to\infty} [a_0, a_1, \ldots, a_n] \tag{9.30}$$

exists for each sequence (9.29). We will soon find an answer to that question.

Let (a_i) be a sequence as in (9.29). The *nth convergent* of the sequence (a_i) is defined to be $\frac{p_n}{q_n} = [a_0, \ldots, a_n]$. A simple calculation shows that

$$\begin{aligned}
\frac{p_0}{q_0} &= \frac{a_0}{1},\\
\frac{p_1}{q_1} &= \frac{a_0a_1 + 1}{a_1},\\
\frac{p_2}{q_2} &= \frac{a_0(a_1 + \frac{1}{a_2}) + 1}{a_1 + \frac{1}{a_2}} = \frac{a_2p_1 + p_0}{a_2q_1 + q_0}.
\end{aligned}$$

Continuing this way, we see that p_n and q_n are polynomials that depend on a_0, a_1, ... a_n. We are now interested in finding the *polynomials* p_n and q_n, so we will regard a_i as indeterminates, assuming that they do not take any specific integer values. After calculating some first p_n and q_n, we may guess that there are general recursion formulae for polynomials p_n and q_n.

Lemma 9.4.3. *For each* $n \geq 2$,

$$p_n = a_np_{n-1} + p_{n-2}, \tag{9.31}$$

$$q_n = a_nq_{n-1} + q_{n-2}. \tag{9.32}$$

hold for each $n \geq 2$.

Proof. By induction on n. The formulae (9.31) and (9.32) are true for $n = 2$, as we saw before. Assume then that they hold for numbers $\{2, \ldots, n\}$. Then,

$$\begin{aligned}
\frac{p_{n+1}}{q_{n+1}} &= [a_0, \ldots, a_{n-1}, a_n, a_{n+1}] = [a_0, \ldots, a_{n-1}, a_n + \frac{1}{a_{n+1}}] \\
&= \frac{(a_n + \frac{1}{a_{n+1}})p_{n-1} + p_{n-2}}{(a_n + \frac{1}{a_{n+1}})q_{n-1} + q_{n-2}} \\
&= \frac{a_{n+1}(a_n p_{n-1} + p_{n-2}) + p_{n-1}}{a_{n+1}(a_n q_{n-1} + q_{n-2}) + q_{n-1}} \\
&= \frac{a_{n+1}p_n + p_{n-1}}{a_{n+1}q_n + q_{n-1}},
\end{aligned}$$

so the formulae (9.31) and (9.32) are valid for each n. □

If a_1, a_2, ... are natural numbers, then it follows by (9.32) that $q_n \geq q_{n-1} + q_{n-2}$ and the inequality $q_n \geq F_n$, where F_n is the *nth Fibonacci number*,[7] can be proved by induction. In the sequel we will use notation p_n and q_n also for the p_n and q_n evaluated at $\{a_0, a_1, \ldots, a_n\}$. The meaning of individual notation p_n or q_n will be clear by the context.

Using the recursion formulae, it is also easy to prove the following lemma:

Lemma 9.4.4. *For any* $n \geq 1$, $p_n q_{n-1} - p_{n-1} q_n = (-1)^{n-1}$.

From the previous lemma it also follows that the convergents $\frac{p_n}{q_n}$ are always in their lowest terms: for if d divides both p_n and q_n, then d also divides 1. Lemma 9.4.4 has even more interesting consequences: By multiplying

$$p_n q_{n-1} - p_{n-1} q_n = (-1)^{n-1} \tag{9.33}$$

by a_n and using the recursion formulae (9.31) and (9.32) once more, we see that

$$p_n q_{n-2} - p_{n-2} q_n = (-1)^n a_n \tag{9.34}$$

whenever $n \geq 2$. Equations (9.33) and (9.34) can be rewritten as

$$\frac{p_n}{q_n} - \frac{p_{n-1}}{q_{n-1}} = \frac{(-1)^{n-1}}{q_n q_{n-1}} \tag{9.35}$$

and

$$\frac{p_n}{q_n} - \frac{p_{n-2}}{q_{n-2}} = \frac{(-1)^n a_n}{q_n q_{n-2}}. \tag{9.36}$$

Equation (9.35) implies inequality

---

[7] Fibonacci numbers F_n are defined by $F_0 = F_1 = 1$ and by $F_n = F_{n-1} + F_{n-2}$.

$$\frac{p_{2n}}{q_{2n}} < \frac{p_{2n-1}}{q_{2n-1}},$$

which, together with (9.36), implies that

$$\frac{p_{2n-2}}{q_{2n-2}} < \frac{p_{2n}}{q_{2n}} < \frac{p_{2n-1}}{q_{2n-1}} < \frac{p_{2n-3}}{q_{2n-3}}. \tag{9.37}$$

Inequalities (9.37) show that the sequence of even convergents is strictly increasing but bounded above by the sequence of odd convergents which is strictly decreasing. It follows by (9.35), and by the fact that q_n tends to infinity (recall that $q_n \geq F_n$), that both of the sequences converge to a limit α. Therefore, each sequence (9.29) represents some irrational number α. Moreover, since the limit α is always between two consecutive convergents, (9.35) implies that

$$\left|\frac{p_n}{q_n} - \alpha\right| \leq \frac{1}{q_n q_{n+1}} < \frac{1}{q_n^2}. \tag{9.38}$$

Example 9.4.3. The continued fraction expansion of π begins with

$$\pi = [3, 7, 15, 1, 292, 1, 1, \ldots].$$

Since $q_4 = 113$ and $q_5 = 292 \cdot q_4 + q_3 = 33102$, $\frac{p_4}{q_4} = \frac{355}{113}$ is a very good approximation of π with a relatively small nominator. By (9.38),

$$\left|\frac{355}{113} - \pi\right| \leq \frac{1}{113 \cdot 33102} = 0.000000267342\ldots.$$

In fact,

$$\left|\frac{355}{113} - \pi\right| = 0.000000266764\ldots,$$

so the estimation given by (9.38) is also quite precise.

Remark 9.4.1. Inequality (9.38) tells us that, for an irrational α, the convergents $\frac{p_n}{q_n}$ give infinitely many rational approximations $\frac{p}{q}$ with $\gcd(p, q) = 1$ of α so precise that

$$\left|\frac{p}{q} - \alpha\right| < \frac{1}{q^2}. \tag{9.39}$$

This has further number theory interest, since the rational numbers have only finitely many rational approximations (9.39) such that $\gcd(p, q) = 1$. This is true because, if $\alpha = \frac{r}{s}$ with $s > 0$, then for $q \geq s$ and $\frac{p}{q} \neq \frac{r}{s}$,

$$\left|\frac{p}{q} - \frac{r}{s}\right| = \left|\frac{ps - qr}{qs}\right| \geq \frac{1}{qs} \geq \frac{1}{q^2}.$$

For $0 < q < s$ there are clearly only finitely many approximations $\frac{p}{q}$ that satisfy (9.39) and $\gcd(p, q) = 1$.

Thus, the irrational numbers can be characterized by the property that they have infinitely many rational approximations (9.38). The continued fraction expansion also gives finitely many rational approximations (9.39) for rational numbers. On the other hand, we will show in a moment (Theorem 9.4.3) that approximations which are good enough are convergents.

It can even be shown that the convergents to a number α are the "best" rational approximations of α in the following sense. For the proof of the following theorem, consult [44].

Theorem 9.4.1. *Let $\frac{p_n}{q_n}$ be a convergent to α. If $0 < q \leq q_n$ and $\frac{p}{q} \neq \frac{p_n}{q_n}$, then*

$$\left|\frac{p_n}{q_n} - \alpha\right| < \left|\frac{p}{q} - \alpha\right|$$

whenever $n \geq 2$.

Now we will show that approximations which are good enough are convergents. To do that, we must first derive an auxiliary result. Let $\alpha = [a_0, a_1, a_2 \ldots]$ be a continued fraction expansion of α. We define $\alpha_{n+1} = [a_{n+1}, a_{n+2}, \ldots]$ and this gives us a formal expression

$$\alpha = [a_0, a_1, a_2, \ldots, a_n, \alpha_{n+1}].$$

This expression is not necessarily a continued fraction in the sense that $\alpha_n \geq 1$ need not be a natural number, but anyway it gives us a representation

$$\alpha = \frac{\alpha_{n+1} p_n + p_{n-1}}{\alpha_{n+1} p_n + q_{n-1}} \tag{9.40}$$

with $p_n q_{n-1} - p_{n-1} q_n = (-1)^{n-1}$ just like in the proof of Lemma 9.4.3. But also any representation looking "enough" like (9.40) implies that $\frac{p_n}{q_n}$ and $\frac{p_{n-1}}{q_{n-1}}$ are convergents to α. More precisely:

Theorem 9.4.2. *Let P, Q, P', and Q' be integers such that $Q > Q' > 0$ and $PQ' - QP' = \pm 1$. If also $\alpha' \geq 1$ and*

$$\alpha = \frac{\alpha' P + P'}{\alpha' Q + Q'},$$

then $\frac{P'}{Q'} = \frac{p_{n-1}}{q_{n-1}}$ and $\frac{P}{Q} = \frac{p_n}{q_n}$ are convergents of a continued fraction expansion of α.

Proof. Anyway, we have a finite continued fraction expansion for

$$\frac{P}{Q} = [a_0, a_1 \ldots, a_n] = \frac{p_n}{q_n}. \tag{9.41}$$

By Lemma 9.4.2, we can choose n to be even or odd, so we can assume, without loss of generality, that

$$PQ' - QP' = (-1)^{n-1}. \tag{9.42}$$

Also by Lemma 9.4.4, we have $\gcd(p_n, q_n) = 1$ and, similarly, $\gcd(P, Q) = 1$. Since Q and q_n have the same signs (Q is positive by assumption), it follows by (9.41) that $P = p_n$ and $Q = q_n$. Using this knowledge and Lemma 9.4.4, equation (9.42) can be rewritten as

$$p_n Q' - q_n P' = (-1)^{n-1} = p_n q_{n-1} - p_{n-1} q_n,$$

or even as

$$p_n(Q' - q_{n-1}) = q_n(P' - p_{n-1}). \tag{9.43}$$

Since $\gcd(p_n, q_n) = 1$, it follows by (9.43) that q_n divides $Q' - q_{n-1}$.

On the other hand, $Q' - q_{n-1} < Q' < q_n$ since $q_n = Q > Q' > 0$ by assumption. Also $q_{n-1} - Q' < q_{n-1} < q_n$ and, by combining these observations, we see that

$$|Q' - q_{n-1}| < q_n,$$

which can be valid only if $Q' = q_{n-1}$. By (9.43) it also follows that $P' = p_{n-1}$. Hence,

$$\alpha = \frac{p_n \alpha' + p_{n-1}}{q_n \alpha' + q_{n-1}},$$

which can also be written as

$$\alpha = [a_0, a_1, \ldots, a_n, \alpha'], \tag{9.44}$$

just like in the proof of Lemma 9.4.3. Since $\alpha' \geq 1$, there is a continued fraction expansion $\alpha' = [a_{n+1}, a_{n+2}, \ldots]$ such that $a_{n+1} \geq 1$. □

Now we are ready to conclude this section by

Theorem 9.4.3. *If*

$$0 < \left| \frac{p}{q} - \alpha \right| \leq \frac{1}{2q^2},$$

then $\frac{p}{q}$ is a convergent to the continued fraction expansion of α.

Proof. By assumption,

$$\frac{p}{q} - \alpha = \sigma\frac{\theta}{q^2},$$

for some $0 < \theta \le \frac{1}{2}$ and $\sigma = \pm 1$. Let $\frac{p_n}{q_n} = [a_0, a_1, \ldots, a_n]$ be the continued fraction expansion for $\frac{p}{q}$. Again by Lemma 9.4.2, we may assume that $\sigma = (-1)^{n-1}$. We define

$$\alpha' = \frac{q_{n-1}}{q_n}\frac{\frac{p_{n-1}}{q_{n-1}} - \alpha}{\alpha - \frac{p_n}{q_n}},$$

where $\frac{p_n}{q_n} = \frac{p}{q}$ and $\frac{p_{n-1}}{q_{n-1}}$ are the last and the second-last convergents to $\frac{p}{q}$. Thus, we have

$$\alpha = \frac{\alpha' p_n + p_{n-1}}{\alpha' q_n + q_{n-1}}$$

and

$$(-1)^{n-1}\frac{\theta}{q_n^2} = \frac{p_n}{q_n} - \alpha = \frac{(-1)^{n-1}}{q_n(\alpha' q_n + q_{n-1})}.$$

Thus,

$$\alpha' = \frac{1}{\theta} - \frac{q_{n-1}}{q_n} > 2 - 1 = 1.$$

By Theorem 9.4.2, $\frac{p_{n-1}}{q_{n-1}}$ and $\frac{p_n}{q_n} = \frac{p}{q}$ are consecutive convergents to α. □

9.5 Shannon Entropy and Information

The purpose of this section is to represent, following [20], the very elementary concepts of information theory. In the beginning, we concentrate on information represented by *binary digits*: bits.

9.5.1 Entropy

By using a single bit we can represent two different configurations: the bit is either 0 or 1. Using two bits, four different configurations are possible, three bits allow eight configurations, etc. Inspired by these initial observations, we say that the *elementary binary entropy* of a set with n elements is $\log_2 n$. The elementary binary entropy, thus, approximately measures the length of bit strings which we need to label all the elements of the set.

Remark 9.5.1. Consider the problem of specifying one single element out of given n elements such that the probability of being the outcome is $\frac{1}{n}$ for each element. The elementary binary entropy $\log_2 n$ thus measures the initial uncertainty in bits, i.e., to specify the outcome, we have to gain $\log_2 n$ bits of information. In other words, we have to obtain the binary string of $\log_2 n$ bits that labels the outcome.

On the other hand, using an alphabet of 3 symbols, we would say that the *elementary ternary uncertainty* of a set of n elements is $\log_3 n$. To get a more unified approach, we define the *elementary entropy* of a set of n elements to be

$$H(n) = K \log n,$$

where K is a constant and log is the natural logarithm. Choosing $K = \frac{1}{\log 2}$ gives the elementary binary information, $K = \frac{1}{\log 3}$ gives the ternary, etc.

However, it feels well justified to say that the information needed in average is less than $K \log n$ if the elements appear as outcomes with non-uniform probability distribution. For example, if we have a priori knowledge that some element does not appear at all, we can say that the information needed, i.e., the initial uncertainty of the outcome, is at most $K \log(n-1)$. How do we deal with a set of n elements, each appearing with a known probability of p_i?.

Assume that we choose l times an element of a set having n elements with uniform probability distribution. This experiment can also be viewed as specifying a string of length l over an n-letter alphabet. There are n^l such strings, and since the letters appear with equal probabilities, the strings do as well. Thus, the elementary entropy of the string set is

$$H(n^l) = K \log n^l = l \cdot K \log n = l \cdot H(n),$$

which is exactly what we could expect: the elementary entropy of strings length l is l times the elementary entropy of a single letter. We use a similar idea to handle the non-uniform probability distributions.

Let us choose l elements of an n-element set having probability distribution $p_1, \ldots, p_n$. Therefore, with large enough l, the ith letter in any such string should appear approximately $k_i = p_i l$ times. Let us study a simplified case where each $k_i = p_i l$ is an integer. A simple combinatorial calculation shows that there are exactly

$$\frac{l!}{k_1! \cdots k_n!}$$

strings where the ith letter occurs $k_i = p_i l$ times. Moreover, the distribution of such strings tends to uniform distribution as l tends to infinity. The elementary entropy of the set of such strings is

$$K \log \frac{l!}{k_1! \cdots k_n!} = K(\log l! - \log k_1! - \ldots - \log k_n!).$$

By using $\log k! = k \log k - k + O(\log k)$ and $l = k_1 + \ldots + k_n$, we can write the average entropy (per letter) as

$$\begin{aligned}
& H(p_1, \ldots, p_n) \\
& = \frac{1}{l} \cdot K\Big(l \log l - l + O(\log l) - \sum_{i=1}^{n} (k_i \log k_i - k_i + O(\log k_i))\Big) \\
& = K\frac{1}{l}\Big(\sum_{i=1}^{n} k_i \log l + O(\log l) - \sum_{i=1}^{n} k_i \log k_i + O(\log l)\Big) \\
& = -K \sum_{i=1}^{n} \frac{k_i}{l} \log \frac{k_i}{l} + O(\frac{\log l}{l}) \\
& = -K \sum_{i=1}^{n} p_i \log p_i + O(\frac{\log l}{l}).
\end{aligned}$$

By letting $l \to \infty$, the last term tends to 0. This encourages the following definition.

Definition 9.5.1. *Suppose that the elements of an n-element set occur with probabilities of $p_1, \ldots, p_n$. The Shannon entropy of distribution $p_1, \ldots, p_n$ is defined to be*

$$H(p_1, \ldots, p_n) = -K(p_1 \log p_1 + \ldots p_n \log p_n).$$

Remark 9.5.2. Because $\lim_{p \to 0} p \log p = 0$, we define $0 \cdot \log 0 = 0$. Clearly, $H(p_1, \ldots, p_n) \geq 0$.

Remark 9.5.3. If $K = \frac{1}{\log 2}$ in the above definition, we talk about *binary Shannon entropy.* The special case that $p_i = \frac{1}{n}$ for each i yields $H(n) = -Kn \cdot \frac{1}{n} \log \frac{1}{n} = K \log n$, again giving the elementary entropy, which we could now call *uniform entropy* as well .

Remark 9.5.4. Recall that the elementary binary entropy measures how many bits of information we have to receive in order to specify an element in a set of n elements with uniform probability distribution. In other words, it measures the uncertainty of outcome in bits. On the other hand, binary Shannon entropy measures the *average* uncertainty in bits in a set with a probability distribution $p_1, \ldots, p_n$.

Since log is a *concave* function,[8] it follows easily that for any probability distribution $p_1, \ldots, p_n$, inequality

[8] A real function f is concave in an interval I if the graph of f is above the chord connecting any points $f(x_1)$ and $f(x_2)$, $x_1, x_2 \in I$. Formally, a concave function f must satisfy $f(\lambda x_1 + (1 - \lambda)x_2) \geq \lambda f(x_1) + (1 - \lambda)f(x_2)$ for $\lambda \in [0, 1]$ and x_1, $x_2 \in I$.

$$p_1 \log x_1 + \ldots + p_n \log x_n \leq \log(p_1 x_1 + \ldots + p_n x_n)$$

holds true. Therefore,

$$\begin{aligned} H(p_1, \ldots, p_n) &= p_1 \log p_1^{-1} + \ldots + p_n \log p_n^{-1} \\ &\leq \log(p_1 p_1^{-1} + \ldots + p_n p_n^{-1}) = \log n. \end{aligned}$$

Notice that the Shannon entropy $H(p_1, \ldots, p_n)$ also attains the above upper bound at $p_1 = \ldots = p_n = \frac{1}{n}$. Moreover, Shannon entropy has the following properties:

1. $H(p_1, \ldots, p_n)$ is a symmetric continuous function.
2. $H(\frac{1}{n}, \ldots, \frac{1}{n})$ is a non-negative, strictly increasing function of variable n.
3. If $(p_1, \ldots, p_n)$ and $(q_1, \ldots, q_m)$ are probability distributions, then

$$\begin{aligned} &H(p_1, \ldots, p_n) + p_n H(q_1, \ldots, q_m) \\ &= H(p_1, \ldots, p_{n-1}, p_n q_1, \ldots, p_n q_m). \end{aligned}$$

We will make the definition of Shannon entropy even more natural by showing that the converse also holds:

Theorem 9.5.1. *If $H(p_1, \ldots, p_n)$ satisfies the conditions 1–3 above, then $H(p_1, \ldots, p_n) = -K(p_1 \log p_1 + \ldots + p_n \log p_n)$ for some positive constant K.*

Proof. We will first prove an auxiliary result that will be of great help: if f is a strictly increasing, non-negative function defined on natural numbers such that $f(s^m) = mf(s)$, then $f(s) = K \log s$ for some positive constant K. To prove this, we take an arbitrary natural number n and fix m such that $s^m \leq 2^n < s^{m+1}$. Then

$$\frac{m}{n} \leq \frac{\log 2}{\log s} < \frac{m}{n} + \frac{1}{n}. \tag{9.45}$$

Because f is strictly increasing, we have also $f(s^m) \leq f(2^n) < f(s^{m+1})$, which implies that

$$\frac{m}{n} \leq \frac{f(2)}{f(s)} < \frac{m}{n} + \frac{1}{n}. \tag{9.46}$$

Inequalities (9.45) and (9.46) together imply that

$$\left| \frac{f(2)}{f(s)} - \frac{\log 2}{\log s} \right| \leq \frac{1}{n}.$$

Since n can be chosen arbitrarily great, we must have $f(s) = \frac{f(2)}{\log 2} \log s = K \log s$. Constant $K = \frac{f(2)}{\log 2}$ is positive, since f is strictly increasing and $f(1) = f(1^0) = 0f(1) = 0$.

For any natural number n, we define $f(n) = H(\frac{1}{n}, \ldots, \frac{1}{n})$. We will demonstrate that $f(s^m) = mf(s)$. Symmetry of H and condition 3 imply that

$$H(\frac{1}{s^m},\dots,\frac{1}{s^m}) = H(\frac{1}{s},\dots,\frac{1}{s}) + s\cdot\frac{1}{s}H(\frac{1}{s^{m-1}},\dots,\frac{1}{s^{m-1}}),$$

which is to say that $f(s^m) = f(s) + f(s^{m-1})$. Claim $f(s^m) = mf(s)$ follows now by induction. Since f is strictly increasing and non-negative by 2, it follows that $f(s) = K\log s$ for some positive constant K.

Let p_i, …, p_n be some rational numbers such that $p_1 + \ldots + p_n = 1$. We can assume that $p_i = \frac{m_i}{N}$, where $N = m_1 + \ldots + m_n$. Using 3 we get

$$\begin{aligned}
&= H(\frac{1}{N},\dots,\frac{1}{N})\\
&= H(\underbrace{p_1\cdot\frac{1}{m_1},\dots,p_1\cdot\frac{1}{m_1}}_{m_1},\underbrace{p_2\cdot\frac{1}{m_2},\dots\dots}_{m_2},\dots,\underbrace{\dots,p_n\cdot\frac{1}{m_n}}_{m_n})\\
&= H(p_1,\dots,p_n) + \sum_{i=1}^{n} p_i H(\frac{1}{m_i},\dots,\frac{1}{m_i}).
\end{aligned}$$

Hence,

$$\begin{aligned}
H(p_1,\dots,p_n) &= -\sum_{i=1}^{n} p_i H(\frac{1}{m_i},\dots,\frac{1}{m_i}) + H(\frac{1}{N},\dots,\frac{1}{N})\\
&= -K(\sum_{i=1}^{n} p_i \log m_i - \log N)\\
&= -K(\sum_{i=1}^{n} p_i \log m_i - \sum_{i=1}^{n} p_i \log N)\\
&= -K\sum_{i=1}^{n} p_i \log p_i,
\end{aligned}$$

which demonstrates that the theorem is true for rational probabilities. Since H is continuous, the result extends to real numbers straightforwardly. □

9.5.2 Information

We will hereafter ignore the constant K in the definition of entropy.

Let $X = \{x_1,\dots,x_n\}$ be a set of n elements, and $p(x_1)$, …, $p(x_n)$ the corresponding probabilities of elements to occur. Similarly, let Y be a set with m elements with a probability distribution $p(y_1)$, …, $p(y_m)$. Such sets are here identified with *discrete random variables*: we regard X as a variable which has n potential values x_1, …, x_n, any such with probability $p(x_i)$.

The *joint entropy* of sets (random variables) X and Y is simply defined as the entropy of set $X \times Y$, where the corresponding probability distribution is the joint distribution $p(x_i, y_j)$:

$$H(X,Y) = -\sum_{i=1}^{n}\sum_{j=1}^{m} p(x_i, y_j) \log p(x_i, y_j).$$

The *conditional entropy* of X, when the value of Y is known to be y_j, is defined by merely replacing the distribution $p(x_i)$ with conditional probability distribution $p(x_i|y_j)$:

$$H(X|y_j) = -\sum_{i=1}^{n} p(x_i|y_j) \log p(x_i|y_j).$$

The *conditional entropy* of X when Y is known is defined as the expected value

$$H(X|Y) = \sum_{j=1}^{m} p(y_j) H(X|y_j).$$

By using the concavity of the function log, one can quite easily prove the following lemma.

Lemma 9.5.1. $H(X|Y) \le H(X)$.

Remark 9.5.5. The above lemma is quite natural: it states that the knowledge of Y cannot increase the uncertainty about X.

Conditional and joint entropy are connected in the following lemma, which has a straightforward proof if one keeps in mind that $p(x_i, y_j) = p(x_i|y_j)p(y_j)$.

Lemma 9.5.2. $H(X|Y) = H(X,Y) - H(Y)$.

Finally, we have all the necessary tools for defining the *information* of X when Y is known.

Definition 9.5.2. $I(X|Y) = H(X) - H(X|Y)$.

Definition (9.5.2) contains the natural idea that the knowledge that Y can provide about X is merely the uncertainty of X minus the uncertainty of X provided Y is known. According to Lemma (9.5.1), $I(X|Y) \ge 0$ always, and trivially $I(X|Y) \le H(X)$.

To end this section, we list some properties of entropy and information. The proofs of the following lemmata are left as exercises.

Lemma 9.5.3. $H(X,Y) \le H(X) + H(Y)$.

Lemma 9.5.4. *Information is symmetric:* $I(X|Y) = I(Y|X)$.

The above lemma justifies the terminology *mutual information of* X and Y.

For the following lemma, we need some extra terminology: if X and Y are random variables, we say that X (randomly) depends on Y if there are conditional probabilities $p(x_i|y_j)$ such that

$$p(x_i) = \sum_{j=1}^{m} p(x_i|y_j) p(y_j).$$

Lemma 9.5.5. *Let X be a random variable that depends on Y, which is a random variable depending on Z. If X is conditionally independent of Z, then $I(Z|X) \leq I(Z|Y)$.*

9.5.3 The Holevo Bound

Let $\rho \in L(H_n)$ be a state (or density matrix) of an n-dimensional quantum system. We know (see Section 8.4.1) that ρ has a spectral representation

$$\rho = \lambda_1 |\boldsymbol{x}_1\rangle\langle\boldsymbol{x}_1| + \ldots + \lambda_n |\boldsymbol{x}_n\rangle\langle\boldsymbol{x}_n|,$$

where each λ_i satisfies $\lambda_i \geq 0$ and

$$\lambda_1 + \ldots + \lambda_n = 1.$$

The *von Neumann entropy* of the state ρ is defined by

$$S(\rho) = -\mathrm{Tr}(\rho \log \rho).$$

According to (8.19), we define $\rho \log \rho$ as

$$\rho \log \rho = \lambda_1 \log \lambda_1 |\boldsymbol{x}_1\rangle\langle\boldsymbol{x}_1| + \ldots + \lambda_n \log \lambda_n |\boldsymbol{x}_n\rangle\langle\boldsymbol{x}_n|,$$

and therefore

$$S(\rho) = -\mathrm{Tr}(\rho \log \rho) = -(\lambda_1 \log \lambda_1 + \ldots + \lambda_n \log \lambda_n).$$

That is to say that the von Neumann entropy of a quantum state $\rho \in L(H_n)$ is exactly the Shannon entropy of distribution $(\lambda_1, \ldots, \lambda_n)$.

Remark 9.5.6. From the definition it follows easily that the von Neumann entropy of a pure state is always 0.

The *Holevo bound* establishes an upper bound of the *accessible information* of quantum systems. For the statement of Holevo's theorem, we assume that there is a source which produces quantum states $\rho_1, \ldots, \rho_n \in L(H_n)$ with probabilities $p_1, \ldots, p_n$. We define X to be the random variable that determines which one of the states ρ_i was produced. That is, the value of X is i if and only if the source produces ρ_i. Let also Y be an observable on H_n. Then the following theorem holds.

Theorem 9.5.2 (The Holevo Bound, [47]).

$$I(X \mid Y) \leq S\Big(\sum_{i=1}^{n} p_i \rho_i\Big) - \sum_{i=1}^{n} p_i S(\rho_i).$$

Remark 9.5.7. By saying that "Y is an observable on H_n" we mean the following: There is a collection $\{E_1, \ldots, E_m\}$ of mutually orthogonal subspaces of H_n such that $H_n = E_1 \oplus \ldots \oplus E_m$ (see Definition 8.3.1), and each subspace E_i has label i. Y is then defined as the random variable which gets value i if and only if the observable $\{E_1, \ldots, E_m\}$ gets value i.

Remark 9.5.8. The above theorem holds also even if Y is allowed to be a POVM rather than a typical observable, defined as in Definition 8.3.1.

9.6 Exercises

1. Prove that a set $S \subseteq V$ is linearly dependent if and only if for some element $\boldsymbol{x} \in S$, $\boldsymbol{x} \in L(S \setminus \{\boldsymbol{x}\})$.
2. Show that if B and B' are two bases of a H_n, then necessarily $|B| = |B'|$.
3. a) Let $n \geq 2$ and $\boldsymbol{x}_1, \boldsymbol{x}_2 \in H_n$. Show that $\boldsymbol{y} \in H_n$ can be chosen in such a way that $L(\boldsymbol{x}_1, \boldsymbol{y}) = L(\boldsymbol{x}_1, \boldsymbol{x}_2)$ and $\langle \boldsymbol{x}_1 \mid \boldsymbol{y} \rangle = 0$.
   b) Generalize a) into a procedure for finding an orthonormal basis of H_n (the procedure is called the *Gram-Schmidt method*).
4. Prove that function $f : \mathbb{N} \to \mathbb{C}$ defined by $f(n) = \frac{1}{n}$ is in $L_2(\mathbb{C})$ but cannot be expressed as a linear combination of $E = \{\boldsymbol{e}_1, \boldsymbol{e}_2, \boldsymbol{e}_3, \ldots\}$.
5. Prove the parallelogram rule: in inner product space V, the equation

$$||\boldsymbol{x} + \boldsymbol{y}||^2 + ||\boldsymbol{x} - \boldsymbol{y}||^2 = 2||\boldsymbol{x}||^2 + 2||\boldsymbol{y}||^2$$

   holds for any $\boldsymbol{x}, \boldsymbol{y} \in V$.
6. Show that the subspace W of $L_2(\mathbb{C})$ generated by vectors $\{\boldsymbol{e}_1, \boldsymbol{e}_2, \boldsymbol{e}_3, \ldots\}$ is not a Hilbert space (cf. Example 9.3.4).
7. Prove Lemma 9.4.1.
8. Prove Lemma 9.4.4.
9. Prove Lemmata 9.5.3 and 9.5.4.
10. Use the properties of the function log to prove Lemma 9.5.5.

References

1. Manindra Agarwal, Neeraj Kayal, and Nitin Saxena: *PRIMES is in P*. Electronically available at http://www.cse.iitk.ac.in/primality.pdf.
2. Andris Ambainis: *A note on quantum black-box complexity of almost all Boolean functions*, Information Processing Letters 71, 5–7 (1999). Electronically available at quant-ph/9811080.[9]
3. Andris Ambainis: *Polynomial degree vs. quantum query complexity*. Electronically available at quant-ph/0305028.
4. E. Bach and J. Shallit: *Computational number theory*, MIT Press (1996).
5. Adriano Barenco, Charles H. Bennett, Richard Cleve, David P. DiVincenzo, Norman Margolus, Peter Shor, Tycho Sleator, John Smolin, and Harald Weinfurter: *Elementary gates for quantum computation*, Physical Review A 52:5, 3457–3467 (1995). Electronically available at quant-ph/9503016.
6. R. Beals, H. Buhrman, R. Cleve, M. Mosca, and R. de Wolf: *Quantum lower bounds by polynomials*, Proceedings of the 39th annual IEEE Symposium on Foundations of Computer Science – FOCS, 352–361 (1998). Electronically available at quant-ph/9802049.
7. P. A. Benioff: *Quantum mechanical Hamiltonian models of discrete processes that erase their own histories: application to Turing machines*, International Journal of Theoretical Physics 21:3/4, 177–202, (1982).
8. Charles H. Bennett: *Logical reversibility of computation*, IBM Journal of Research and Development 17, 525–532 (1973).
9. Charles H. Bennett: *Time/space trade-offs for reversible computation*, SIAM Journal of Computing 18, 766–776 (1989).
10. Charles H. Bennett, Ethan Bernstein, Gilles Brassard, and Umesh V. Vazirani: *Strengths and weaknesses of quantum computation*, SIAM Journal of Computing 26:5, 1510–1523 (1997). Electronically available at quant-ph/9701001.
11. Charles H. Bennett, Gilles Brassard, Claude Crépeau, Richard Jozsa, Ahsher Peres, Williams K. Wootters: *Teleporting an unknown quantum state via dual classical and Einstein-Podolky-Rosen channels*. Physical Rewiev Letters 70, 1895–1899 (1993).
12. Charles H. Bennett, Stephen J. Wiesner: *Communication via one- and two-particle operators on Einstein-Podolsky-Rosen states*. Physical Review Letters, 69(20): 2881–2884 (1992).
13. Ethan Bernstein and Umesh Vazirani: *Quantum complexity theory*, SIAM Journal of Computing 26:5, 1411–1473 (1997).
14. André Berthiaume and Gilles Brassard: *Oracle quantum computing*, Proceedings of the Workshop on Physics and Computation – PhysComp'92, IEEE Press, 195–199 (1992).

---

[9] Code "quant-ph/9811080" refers to http://xxx.lanl.gov/abs/quant-ph/9811080 at Los Alamos preprint archive.

15. D. Boschi, S. Branca, F. De Martini, L. Hardy, S. Popescu: *Experimental realization of teleporting an unknown pure quantum state via dual classical and Einstein-Podolsky-Rosen channels* Physical Review Letters 80:6, 1121–1125 (1998). Electronically available at quant-ph/9710013.
16. Dik Bouwmeester, Jian-Wei Pan, Klaus Mattle, Manfred Eibl, Harald Weinfurter, Anton Zeilinger: *Experimental quantum teleportation*, Nature 390, 575–579 (1997).
17. Michel Boyer, Gilles Brassard, Peter Høyer, and Alain Tapp: *Tight bound on quantum searching*, Fourth Workshop on Physics and Computation – PhysComp'96, Ed: T. Toffoli, M. Biaford, J. Lean, New England Complex Systems Institute, 36–43 (1996). Electronically available at quant-ph/9605034.
18. Gilles Brassard and Peter Høyer: *An exact quantum polynomial-time algorithm for Simon's problem*, Proceedings of the 1997 Israeli Symposium on Theory of Computing and Systems – ISTCS'97, 12–23 (1997). Electronically available at quant-ph/9704027.
19. Gilles Brassard, Peter Høyer, and Alain Tapp: *Quantum counting*, Automata, Languages and Programming, Proceedings of the 25th International Colloquium, ICALP'98, Lecture Notes in Computer Science 1443, 820–831, Springer (1998). Electronically available at quant-ph/9805082.
20. Leon Brillouin: *Science and information theory*, 2nd edition, Academic Press (1967).
21. Harry Buhrman and Wim van Dam: *Quantum Bounded Query Complexity*, Proceedings of the 14th Annual IEEE Conference on Computational Complexity – CoCo'99, 149–157 (1999). Electronically available at quant-ph/9903035.
22. Paul Busch: *Quantum states and generalized observables: a simple proof of Gleason's theorem.* Electronically available at quant-ph/9909073.
23. Paul Busch, Pekka J. Lahti, and P. Mittelstaedt: *The quantum theory of measurement*, Springer-Verlag, 1991.
24. A. R. Calderbank, Peter W. Shor: *Good quantum error-correcting codes exist*, Physical Review A 54, 1098–1105 (1996). Electronically available at quant-ph/9512032.
25. Man-Duen Choi: *Completely positive linear maps on complex matrices*, Linear Algebra and its Applications 10, 285–290 (1975).
26. Isaac L. Chuang, Lieven M. K. Vandersypen, Xinlan Zhou, Debbie W. Leung, and Seth Lloyd: *Experimental realization of a quantum algorithm*, Nature 393, 143–146 (1998). Electronically available at quant-ph/9801037.
27. Juan I. Cirac and Peter Zoller: *Quantum computations with cold trapped ions*, Physical Review Letters 74:20, 4091–4094 (1995).
28. Henri Cohen: *A course in computational algebraic number theory*, Graduate Texts in Mathematics 138, Springer (1993), 4th printing 2000.
29. Win van Dam: *Two classical queries versus one quantum query*. Electronically available at quant-ph/9806090.
30. David Deutsch: *Uncertainty in quantum measurements*, Physical Review Letters 50:9, 631–633 (1983).
31. David Deutsch: *Quantum theory, the Church-Turing principle and the universal quantum computer*, Proceedings of the Royal Society of London A 400, 97–117 (1985).
32. David Deutsch: *Quantum computational networks*, Proceedings of the Royal Society of London A 425, 73–90 (1989).
33. David Deutsch, Adriano Barenco, and Artur Ekert: *Universality in quantum computation*, Proceedings of the Royal Society of London A 449, 669–677 (1995). Electronically available at quant-ph/9505018.

34. David Deutsch, Richard Jozsa: *Rapid solutions of problems by quantum computation*, Proceedings of the Royal Society of London A 439, 553–558 (1992).
35. Chistoph Dürr and Peter Høyer: *A quantum algorithm for finding the minimum.* Electronically available at quant-ph/9607014.
36. Mark Ettinger and Peter Høyer: *On quantum algorithms for noncommutative hidden subgroups*, Proceedings of the 16th Annual Symposium on Theoretical Aspects of Computer Science – STACS 99, Lecture Notes in Computer Science 1563, 478–487, Springer (1999). Electronically available at quant-ph/9807029.
37. Edward Farhi, Jeffrey Goldstone, Sam Gutmann, and Michael Sipser: *A limit on the speed of quantum computation in determining parity*, Physical Review Letters 81:5, 5442–5444 (1998). Electronically available at quant-ph/9802045.
38. Richard P. Feynman: *Simulating physics with computers*, International Journal of Theoretical Physics 21:6/7, 467–488 (1982).
39. A. Furusawa, J. L. Sorensen, S. L. Braunstein, C. A. Fuchs, H. J. Kimble, E. S. Polzik: *Unconditional quantum teleportation*, Science 282, 706–709 (1998).
40. Daniel Gottesmann: *The Heisenberg Representation of Quantum Computers.* Electronically available at quant-ph/9807006.
41. Lov K. Grover: *A fast quantum-mechanical algorithm for database search*, Proceedings of the 28th Annual ACM Symposium on the Theory of Computing – STOC, 212–219 (1996). Electronically available at quant-ph/9605043.
42. Lov K. Grover and Terry Rudolph: *How significant are the known collision and element distinctness quantum algorithms?* Electronically available at quant-ph/0306017.
43. Josef Gruska: *Quantum Computing*, McGraw-Hill (1999).
44. G. H. Hardy and E. M. Wright: *An introduction to the theory of numbers*, 4th ed with corrections, Clarendon Press, Oxford (1971).
45. Mika Hirvensalo: *On quantum computation*, Ph.Lic. Thesis, University of Turku, 1997.
46. Mika Hirvensalo: *The reversibility in quantum computation theory*, Proceedings of the 3rd International Conference Developments in Language Theory – DLT'97, Ed.: Symeon Bozapalidis, Aristotle University of Thessaloniki, 203–210 (1997).
47. A. S. Holevo: *Statistical Problems in Quantum Physics*, Proceedings of the Second Japan-USSR Symposium on Probability Theory, Eds.: G. Murayama and J.V. Prokhorov, Springer, 104–109 (1973).
48. Richard Jozsa: *A stronger no-cloning theorem.* Electronically available at quant-ph/0204153.
49. Loo Keng Hua: *Introduction to number theory*, Springer-Verlag, 1982.
50. A. Y. Kitaev: *Quantum computation: algorithms and error correction*, Russian Mathematical surveys 52:1991 (1997).
51. E. Knill, R. Laflamme, R. Martinez, C.-H. Tseng: *An algorithmic benchmark for quantum information processing*, Nature 404: 368–370 (2000).
52. Rolf Landauer: *Irreversibility and heat generation in the computing process*, IBM Journal of Research and Development 5, 183-191 (1961).
53. M. Y. Lecerf: *Récursive insolubilité de l'équation générale de diagonalisation de deux monomorphismes de monoïdes libres* $\varphi x = \psi x$, Comptes Rendus de l'Académie des Sciences 257, 2940–2943 (1963).
54. Ming Li, John Tromp and Paul Vitányi: *Reversible simulation of irreversible computation*, Physica D 120:1/2, 168-176 (1998). Electronically available at quant-ph/9703009.
55. Seth Lloyd: *A potentially realizable quantum computer*, Science 261, 1569–1571 (1993).

56. Hans Maassen and J. B. M. Uffink: *Generalized entropic uncertainty relations*, Physical Review Letters 60:12, 1103–1106 (1988).
57. F. J. MacWilliams and Neil J. A. Sloane: *The theory of error-correcting codes*, North-Holland (1981).
58. Gary L. Miller: *Riemann's hypothesis and tests for primality*, Journal of Computer and System Sciences 13, 300–317 (1976).
59. Michele Mosca and Artur Ekert: *The hidden subgroup problem and eigenvalue estimation on a quantum computer*, Quantum Computing and Quantum Communications, Proceedings of the 1st NASA International Conference, Lecture Notes in Computer Science 1509, 174–188, Springer (1998). Electronically available at quant-ph/9903071.
60. A. J. Menezes, P. C. van Oorschot, and S. A. Vanstone: *Handbook of applied cryptography*, CRC Press Series on Discrete and Mathematics and Its Applications, CRC Press (1997).
61. John von Neumann: *Mathematical foundations of quantum mechanics*, Princeton university press, translated from the German edition by Robert T. Beyer (1955)
62. Michael A. Nielsen and Isaac L. Chuang: *Quantum Computation and Quantum Information*, Cambridge University Press (2000).
63. Masanao Ozawa: *Quantum Turing machines: local transitions, preparation, measurement, and halting problem*, Quantum Communication, Computing, and Measurement 2, Eds.: Prem Kumar, G. Mauro D'Ariano, and Osamu Hirota, Kluwer, New York, 241–248 (2000). Electronically available at quant-ph/9809038.
64. Christos H. Papadimitriou: *Computational complexity*, Addison-Wesley (1994).
65. K. R. Parthasarathy: *An introduction to quantum stochastic calculus*, Birkhäuser, Basel (1992).
66. A.K. Pati, S. L. Braunstein: *Impossibility of deleting an unknown quantum state*, Nature 404, 164–165 (2000).
67. R. Paturi: *On the degreee of polynomials that approximate symmetric Boolean functions*, Proceedings of the 28th Annual ACM Symposium on the Theory of Computing - STOC, 468-474 (1992).
68. Max Planck: Annalen der Physik 1, 69, 1900; Verhandlg. dtsch. phys. Ges., 2, 202; Verhandlg. dtsch. phys. Ges. 2, 237; Annalen der Physik 4, 553, 1901.
69. M. B. Plenio and P. L. Knight: *Realistic lower bounds for the factorization time of large numbers on a quantum computer*, Physical Review A 53:5, 2986–2990 (1996). Electronically available at quant-ph/9512001.
70. E. L. Post: *The two-valued iterative systems of mathematical logic*, Princeton University Press (1941).
71. E. L. Post: *A variant of a recursively unsolvable problem*, Bulletin of the American Mathematical Society 52, 264–268 (1946).
72. John Preskill: *Robust solutions to hard problems*, Nature 391, 631–632 (1998).
73. Marcel Riesz: *Sur les maxima des formes bilinéaires et sur les fonctionnelles linéaires*, Acta Mathematica 49, 465–497 (1926).
74. Yurii Roghozin *On the notion of universality and small universal Turing machines*, Theoretical Computer Science 168, 215–240 (1996).
75. Sheldon M. Ross: *Introduction to probability models*, 4th edition, Academic Press (1985).
76. J. Barkley Rosser and Lowell Schoenfeld: *Approximate formulas for some functions of prime numbers*, Illinois Journal of Mathematics 6:1, 64–94 (1962).
77. Walter Rudin: *Functional Analysis*, 2nd edition, McGraw-Hill (1991).

78. Keijo Ruohonen: *Reversible machines and Post's correspondence problem for biprefix morphisms*, EIK – Journal of Information Processing and Cybernetics 21:12, 579–595 (1985).
79. Arto Salomaa: *Public-key cryptography*, Texts in Theoretical Computer Science – An EATCS Series, 2nd ed., Springer (1996).
80. Yaoyun Shi: *Both Toffoli and controlled-NOT need little help to do universal quantum computation.* Quantum Information and Computation 3:1, 84–92 (2003). Electronically available at quant-ph/0205115.
81. Peter W. Shor: *Algorithms for quantum computation: discrete log and factoring*, Proceedings of the 35th annual IEEE Symposium on Foundations of Computer Science – FOCS, 20–22 (1994).
82. Peter W. Shor: *Scheme for reducing decoherence in quantum computer memory*, Physical Review A 52:4, 2493–2496 (1995).
83. Uwe Schöning: *A Probabilistic algorithm for k -SAT based on limited local search and restart*, Algorithmica 32, 615–623 (2002).
84. Daniel R. Simon: *On the power of quantum computation*, Proceedings of the 35th annual IEEE Symposium on Foundations of Computer Science – FOCS, 116–123 (1994).
85. Douglas R. Stinson: *Cryptography - Theory and practice*, CRC Press Series on Discrete Mathematics and Its Applications, CRC Press, Boca Raton (1995).
86. W. Tittel, J. Brendel, H. Zbinden, N. Gisin: *Violation of Bell inequalities by photons more than 10 km apart*, Physical Review Letters 81:17, 3563–3566, (1998). Electronically available at quant-ph/9806043.
87. Tommaso Toffoli: *Bicontinuous extensions of invertible combinatorial functions*, Mathematical Systems Theory 14, 13–23 (1981).
88. B. L. van der Waerden: *Sources of quantum mechanics*, North-Holland (1967).
89. C. P. Williams and S. H. Clearwater: *Explorations in quantum computing*, Springer (1998).
90. C. P. Williams and S. H. Clearwater: *Ultimate zero and one. Computing at the quantum frontier*, Springer (2000).
91. William K. Wootters, Wojciech H. Zurek: *A single quantum cannot be cloned*, Nature 299, 802–803 (1982).
92. Andrew Chi-Chih Yao: *Quantum circuit complexity*, Proceedings of the 34th annual IEEE Symposium on Foundations of Computer Science – FOCS, 352–361 (1993).

Index

Natural Computing Series

W.M. Spears: **Evolutionary Algorithms. The Role of Mutation and Recombination.**
XIV, 222 pages, 55 figs., 23 tables. 2000

H.-G. Beyer: **The Theory of Evolution Strategies.**
XIX, 380 pages, 52 figs., 9 tables. 2001

L. Kallel, B. Naudts, A. Rogers (Eds.): **Theoretical Aspects of Evolutionary Computing.**
X, 497 pages. 2001

G. Păun: **Membrane Computing. An Introduction.**
XI, 429 pages, 37 figs., 5 tables. 2002

A.A. Freitas: **Data Mining and Knowledge Discovery with Evolutionary Algorithms.**
XIV, 264 pages, 74 figs., 10 tables. 2002

H.-P. Schwefel, I. Wegener, K. Weinert (Eds.): **Advances in Computational Intelligence.**
VIII, 325 pages. 2003

A. Ghosh, S. Tsutsui (Eds.): **Advances in Evolutionary Computing.**
XVI, 1006 pages. 2003

L.F. Landweber, E. Winfree (Eds.): **Evolution as Computation.**
DIMACS Workshop, Princeton, January 1999. XV, 332 pages. 2002

M. Amos: **Theoretical and Experimental DNA Computation.**
Approx. 200 pages. 2004

M. Hirvensalo: **Quantum Computing.**
2nd ed., XIII, 214 pages. 2004 (first edition published in the series)

A.E. Eiben, J.E. Smith: **Introduction to Evolutionary Computing.**
XV, 299 pages. 2003

G. Ciobanu (Ed.): **Modelling in Molecular Biology.**
Approx. 300 pages. 2004

A. Ehrenfeucht, T. Harju, I. Petre, D.M. Prescott, G. Rozenberg:
Computation in Living Cells.
Approx. 175 pages. 2004

R. Paton, H. Bolouri, M. Holcombe, J. H. Parish, R. Tateson (Eds.):
Computation in Cells and Tissues.
Approx. 350 pages. 2004

L. Sekanina: **From Theory to Hardware Implementations.**
XVI, 194 pages. 2004